CREATION of the UNIVERSE

CREATION of the UNIVERSE

Fang Li Zhi
Beijing Astronomical Observatory
Chinese Academy of Sciences

Li Shu Xian
Department of Physics
Beijing University

Translated by
T. Kiang
Dunsink Observatory
Dublin

World Scientific
Singapore • New Jersey • London • Hong Kong

Published by

World Scientific Publishing Co. Pte. Ltd.
P O Box 128, Farrer Road, Singapore 9128
USA office: Suite 1B, 1060 Main Street, River Edge, NJ 07661
UK office: 73 Lynton Mead, Totteridge, London N20 8DH

Library of Congress Cataloging-in-Publication Data

Fang, Li Zhi.
Creation of the universe.
1. Cosmology. 2. Astrophysics. I. Li, Shu Xian. II. Title.
QB981.F17 1989 523.1 88-33962
ISBN 9971-50-600-9
ISBN 9971-50-601-7 (pbk.)

First published: 1989
First reprint: 1993

Printed in Singapore by Utopia Press.

To G. Borner, Hong-Yee Chiu, T. Kiang, R. Ruffini,
H. Sato, Orville Schell, J. A. Wheeler

PREFACE

The birth of the universe as a whole is now a challenging problem for physics. The evolution of the universe from nothing is described by the big bang theory. *Creation of the Universe* is the story of advances in cosmology and how they were made. It traces the developments of the big bang theory, from the expansion of the universe to quantum cosmology, from the formation of large scale structures to the physics of the Planck Era. It is also the story of the scientists involved in the search, their frustrations, hardships, hopes and joys when great discoveries are made. The theme of *Creation of the Universe* is similar to an earlier book *From Newton's Laws to Einstein's Theory of Relativity* by Fang Li Zhi and Chu Yao Quan, namely, the unification of the universe. In the book *From Newton's Laws to Einstein's Theory of Relativity* we emphasized the unification of laws which are obeyed by everything. In *Creation of the Universe* we emphase the unification of the origin of everything.

As with the former book, in *Creation of the Universe* we avoid as much as possible the use of advanced mathematics. Instead we present theories and observations of cosmology in a way that the reader can easily understand the basic concepts related to global problems in cosmology.

We wish to, at this point, mention the trouble we had at the beginning of 1987. At that time one of us (Fang Li Zhi) was forced to move from the position of vice-president of the University of Sciences and Technology of China to the Beijing Astronomical Observatory because of the accusation that Fang Li Zhi had stirred up and catalyzed the student's demonstrations

for greater freedom and more democracy in China. As a consequence, this book was almost cancelled in its Chinese edition. However, unfaltering and loyal support from many colleagues finally allowed the Chinese edition to be published according to schedule in 1987. We are very grateful to all those people who made contributions in terms of moral support as mentioned above.

Fang Li Zhi
Li Shu Xian

Beijing, 18th November 1988

TABLE OF CONTENTS

Chapter One

PHYSICAL INTIMATIONS OF AN ANCIENT CHINESE STORY

A Timeless Regret

There is much to be proud of in the thousands of years of Chinese civilization. There are some things, however, which we regret. One of them is our attitude towards the story of the worried men of Qi.

This simple story is very well-known. It just says that once upon a time, the men of the state of Qi were so engrossed in thinking when the sky would fall and when the Earth would collapse that they forgot to sleep or eat. The story originated in the Tian Rui chapter of *Lie Zi* ("[Writings of] Miscellaneous Masters"). The text runs as follows:

> *In the state of Qi, some people worried that the sky and earth might fall and collapse and they might no longer have any support; they worried so much that they neglected their sleep and food.*

From the tone, which is rather neutral, it is impossible to say whether the writer is praising or blaming the men of Qi.

Unfortunately, the phrase *Men of Qi Worrying about the Sky* has since acquired a negative connotation, which is used to describe the thinking of meaningless problems.

More unfortunately, when advances in science today have incontrovertibly proven the great value of the men of Qi's heavenly worry, there are still many people equating the consideration of the sky falling to meaninglessness. We have to admit that this is a thousand-year-old regret.

This regretful thing that happened in China has its cultural background. The poet Li Bai could have, in one of his poems, freely imagined and written, *"I thought it was the Silver River (the Milky Way) falling from the Nine Heavens,"* yet, disregarding the scientific question as to whether the Milky Way can really fall from the heavens, he did not break from the traditional values and wrote:

> *Unappreciated, my loyalty was likened to the groundless worries of the men of Qi.*

It seems that putting a bold and imaginative question in the mouth of a man of letters is *highly romantic*, but in science it becomes a *groundless worry* of no value.

This is not right. One of the aims of this book is to show how the serious, scientific investigation of such a *groundless worry* has contributed and is contributing to the development of the civilization of mankind.

Let us then begin with the question of the sky falling.

Classical Interpretation of the Problem of the Sky Falling

Can the *sky* fall?

This question, even when judged by today's scientific standard, qualifies as a very clearly formulated physical question. If every unsupported object near the Earth's surface will fall to Earth, will things from the sky also fall? Since this is such a natural question to ask, it was considered not only by the men of Qi in ancient China, but also by the philosophers of ancient Greece.

Aristotle believed that the sky or the things in the sky will never fall to the Earth. His reason was that objects below the Moon belong to this world, those above the Moon belong to the gods, and that they are things of different kinds, obeying different laws of movement. The former always fall, the latter always move in circular orbits and will never fall. This is the famous theory of the two regions. It can satisfactorily explain why falling objects on Earth always tend towards the centre of the Earth and why heavenly bodies always revolve around the Earth.

Newton also thought about this problem: if ripe apples always fall, why not the Moon? Newton did not subscribe to Aristotle's idea of the two regions and believed that apples and the Moon should obey the same law. According to Newton's theory, apples and the Moon are both attracted by the Earth. In fact, like apples, the Moon is constantly falling towards the Earth, except

Fig. 1.1. In the state of Qi, some people worried that the sky and earth might fall and collapse and they might no longer have any support; they worried so much that they neglected their sleep and food.

that the Moon has a considerable initial transverse velocity. Thus a falling movement plus a cross movement results in a circle that will never touch the surface of the Earth (Fig. 1.2).

In summary, the reason the Moon does not fall is not because it belongs to some heavenly region, but because of its large initial motion. Why is there a large initial velocity? Newton says this initial velocity was given by the First Mover.

The First Move

The First Move was not invented by Newton; it existed in Aristotle's *Physics*. Aristotle used it to represent the ultimate cause. If we apply the idea of causality to the study of motion of objects, then the motion of a body is the result of its being moved by some other body, or, *any moved body is*

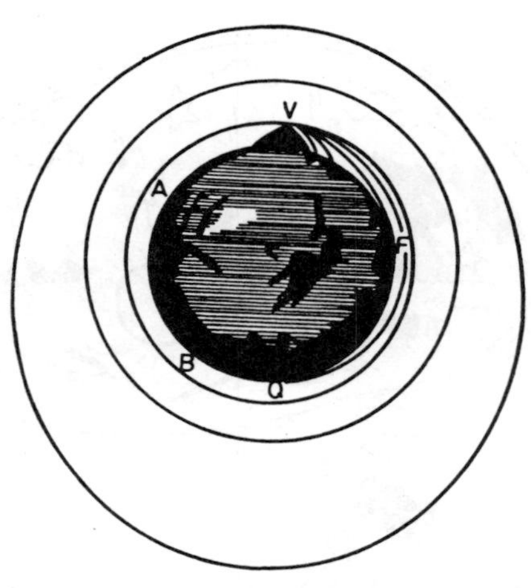

Fig. 1.2. Newton's explanation of the problem of the sky falling. The movement of composing a falling movement and a cross movement is parabolic. When the speed of the cross movement is large enough, the parabolic movement becomes a circle that will never touch the surface of the earth.

moved by some body. Starting from this proposition, we conclude the existence of a *First Mover that is moved by no other body* as the ultimate cause.

Later, during the Middle Ages, the idea of First Mover was used by the theologian Thomas Aquinas as one of the five proofs of the existence of God, the so-called *cosmological proof.* In his "Theological Compendium" *Ser. 1. Pt. 1*, Aquinas wrote:

> *In the world, some objects are in motion, this is clear from our senses and also is certain. All motions result from being moved by some other objects.*
>
> *If an object itself is moving and is necessarily being moved by some other object, then the latter must necessarily be moved by something else. But we certainly cannot repeat this argument indefinitely. Therefore, we eventually arrive at a First Mover who is not moved by anything. This is necessarily so. Everybody knows this First Mover is God.*

Since then, *First Mover* seems to have become a synonym for *God.* Although Newton's solution of the problem of the sky falling was a great step forward from Aristotle's, he still kept and used the notion of the First Mover. This last point has not generally been accepted by researchers in science.

Kant's Evolutionary View

To overcome the difficulty of the First Move, Kant proposed the view that all heavenly bodies are formed through gradual evolution and have not been as they are since the genesis of the universe. This is the evolutionary view.

The notion of the evolution of heavenly bodies plays an important role in the development of science in the West. The notion of an eternal, unchanging universe is much more deeply ingrained in the West than in China. In the times of Kant, it indeed required courage to challenge the eternal view with the evolutionary view, whereas in the Chinese tradition there was no lack of evolutionary views. For example, Yang Quan from the times of the Three Kingdoms (3rd Century A.D.) wrote in his *Wu Li Lun* ("Physics"):

> *Qi (breath) develops and rises, jing hua (essence) floats aloft, turning about following the flow, it is called the Heavenly River (the Milky Way), also called "Yun Han", and all the stars appear.*

This view is very similar to Kant's notion of stars being formed by the contraction of gas clouds.

It is regrettable that the ancient Chinese scholars did not make further use of this masterly insight by applying it to some particular problem. Hence, this notion remained at a philosophical level and did not develop systematically into science. By contrast, a direct motive of Kant's evolutionary proposal is his attempt to solve the question of the Moon's large transverse velocity.

According to Kant, the solar system was formed from a gaseous nebula. Originally the nebula was very large but it contracted because of self-gravitation and eventually condensed into several planets, satellites and the sun. During the process of contraction, matter can acquire very large velocities, just like objects falling from the sky acquire large velocities when reaching the ground. The Moon's large velocity is thus a very natural consequence. However, all the velocities resulting from contraction are directed towards the centre, and not in a transverse direction. Hence we need a mechanism that will convert a high centripetal velocity into a high transverse velocity in order to explain the motion of the Moon.

Kant further hypothesized the existence of a universal repulsive force. The obstruction of this repulsive force changes motion towards a centre into sideways motion. In the language of modern physics, this is the mechanism of scattering or the mechanism of collision: when high-velocity particles encounter a strong repulsive force, their direction of motion can be significantly changed (Fig. 1.3).

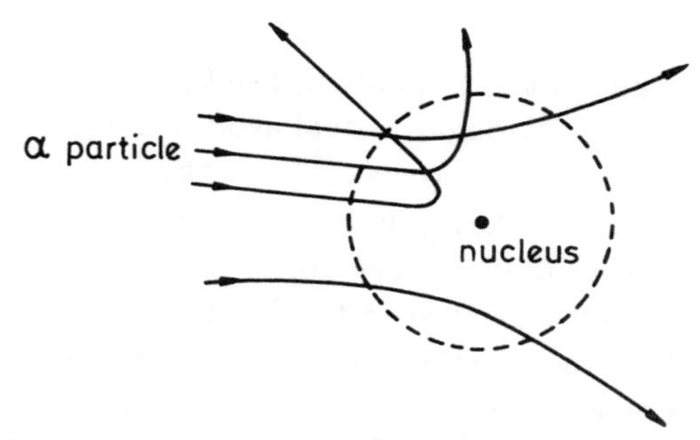

Fig. 1.3. During the collision between a high speed particle and a hard repelling force, a particle will change its direction of movement. For instance, a positively charged α-particle will change its direction by the scattering of a nucleus.

It should be stated that Kant's proposal of the existence of a repulsive force was, to a considerable extent, based on his philosophical view. He had said that attraction and repulsion are equally certain, equally simple, equally basic and general. When matter is resolved into tiny bits, the mutual repulsion is shown in the motion generated by the conflict between repulsion and attraction. In the present context, we can indeed see this philosophical notion of *conflict between repulsion and attraction* at work.

However, deductions based on such philosophical views may not always be correct.

The Ratio 137438953471/137438953472

Laplace gave the first convincing proof that Kant's conjecture based on the above philosophical notion was wrong.

Laplace pointed out that if the cross motions of heavenly bodies were caused by repulsive scattering, then some should be directed to the right or to the left, that is, every direction should have an equal probability. The bodies of the solar system should then be such that some revolve from East to West and some from West to East. But the fact is that,

> ... *all the planets revolve in their orbits around the Sun in the same direction and all the orbits are in the same plane. All satellites also revolve around the planets in the same direction and in the same plane.*

This is an extract from Laplace's *An Outline of Cosmic Systems.* At the time of Laplace, a total of thirty-seven revolving systems were known in the solar system, none of which had a movement from East to West (Fig. 1.4).

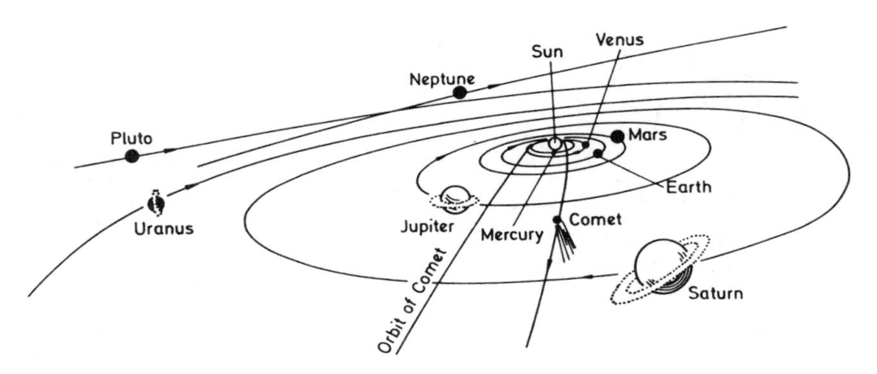

Fig. 1.4. The solar system. As mentioned by Laplace, all the orbits of planets are in the same plane. Only some orbits depart from the plane at small angles, but this does not affect Laplace's argument.

Later, Laplace gave a more quantitative proof. If the solar system is formed in accordance with Kant's mechanism, then the transverse motion of a heavenly body has a probability 1/2 of being from West to East, and a probability 1/2 of being from East to West. The probability that all 37 bodies have eastward cross motion is then $(1/2)^{37}$. Hence Laplace concluded:

If the orbital planes are oriented at random, then the probability that at least 1 out of 37 bodies should move in the retrograde sense is

$$1-\left(\frac{1}{2}\right)^{37}=\frac{137438953471}{137438953472}\,.$$

This shows that Kant's notion of repulsion is wrong at a level of $1-(1/2)^{37}=$ 99.999%.

Furthermore, Laplace showed that, in order to generate large transverse velocities, we do not need repulsion; a universal gravitation suffices. And the velocities so generated will have the same direction, that is to say, the contraction of the gaseous nebula not only increases the centripetal velocity of the particles, it also increases their transverse velocity, and the final system must be one with all components revolving in the same direction. The basis of this conclusion is that with systems under universal gravitation, angular momentum is conserved; as the nebula contracts, its angular momentum remains the same. For a gaseous nebula initially possessing a certain amount of angular momentum, the mere existence of the Law of Universal Gravitation is sufficient to ensure that the system will never contract to a point, and that it will necessarily form a rotating disk structure (Fig. 1.5).

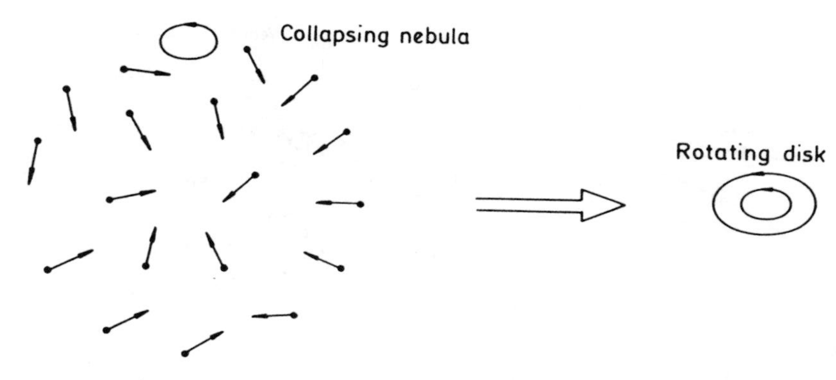

Fig. 1.5. The primeval clouds will only collapse into objects with disc structure if the clouds possess angular momentum.

In brief, the reason the Moon does not fall is precisely because of the existence of universal gravitation which endeavours to make it fall.

Examination of Larger Systems

Laplace's explanation is eminently successful. Starting from a unified viewpoint and law, it explains not only why the bodies of the solar system do not fall, it also explains why these bodies have a disk-like distribution and why they all move in the same direction.

Laplace's theory can also explain larger heavenly systems. It has indeed been discovered that many heavenly systems of various scales have a rotating disk structure.

Viewed astrophysically, the solar system is only a very small disk. Measured by the propagation of light, the radius of the solar system is about one light-hour, i.e., the distance covered by light in 1 hour. Similar disk distributions of matter are present around many stars.

The Milky Way system is a system that includes roughly 10^{11} stars, including the Sun. The Milky Way system is also a rotating disk. Its scale is, of course, much larger; the distance from the Sun to the centre of rotation of the Milky Way is about 10 kiloparsecs or 3×10^4 light-years (Fig. 1.6).

Many galaxies (systems beyond the Milky Way) are called spiral galaxies, precisely because they have a disk-like shape and are rotating. These rotating spirals are each a system of stars similar to the Milky Way system (Fig. 1.7).

Still larger cosmic systems are clusters of galaxies. Each cluster is made up of many galaxies, the larger ones having thousands of constituents. In general, the clusters no longer show a disk shape, but many among them still

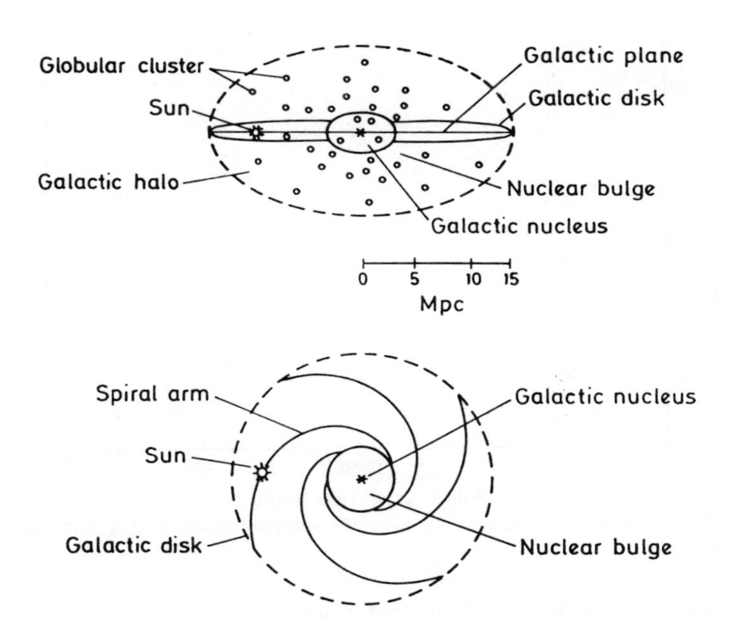

Fig. 1.6. (a) A side view of the Milky Way; (b) A top view of the Milky Way.

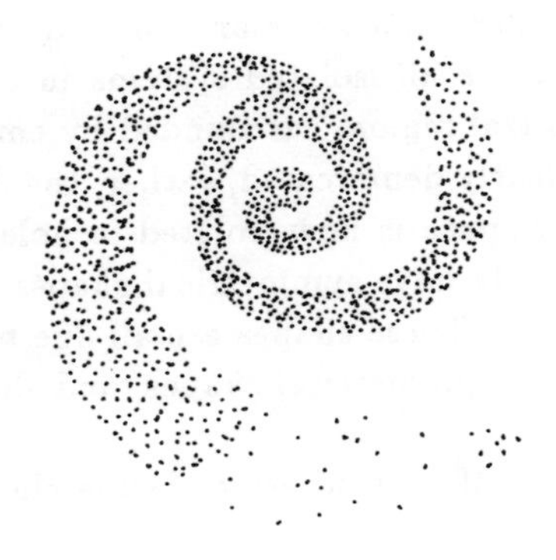

Fig. 1.7. A spiral galaxy.

have rather symmetrical shapes indicating rotation. The scale of a cluster of galaxies is a further factor of several hundred up on that of a galaxy, being on the order of 5×10^6 parsecs or $\sim 2\times10^7$ light-years (Fig. 1.8).

If we go to a still larger scale of $(60\sim100)\times10^6$ parsecs or 2×10^8 light-years, then we come to the scale of superclusters. A supercluster is made of

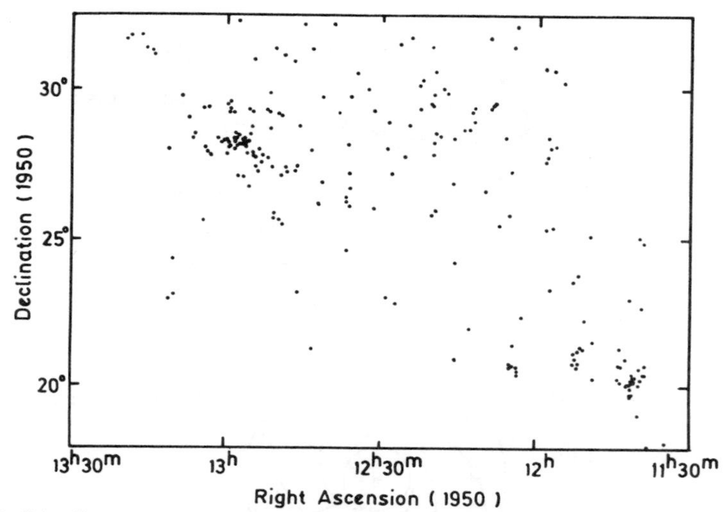

Fig. 1.8. Distribution of galaxies in the area of clusters of Coma and A1367, where a point denotes a galaxy.

many clusters. A number of examples of superclusters are given in Figs. 1.9, 1.10 and 1.11.

From these figures we can see that the shape of superclusters is very different from those of smaller-scale systems. The Sun, the Milky Way, galaxies and clusters of galaxies are all isolated systems in space, that is, matter is concentrated in the central region, surrounded by empty space. In contrast, superclusters are not clearly demarcated, rather, the distribution of matter is net-like, and the empty space is now isolated in holes or "voids". Also, the smaller-scale bodies mostly have symmetrical shapes, whether that of a disk or that of a flattened ball. These shapes are all due to rotation. In contrast, superclusters do not have symmetrical shapes, and show no clear evidence of rotation.

How are we to explain this peculiarity of superclusters?

Time-Scale of the Collapse

The characteristic features of superclusters can also be explained in terms of Laplace's nebular theory. Let us repeat that this theory says that, as long as the original gas cloud has a certain amount of angular momentum, then no matter what its original shape, the body formed after the collapse must have a symmetrical shape indicating rotation. These bodies are stable and there will no longer be any *sky falling*. Note, here we have stressed *after the*

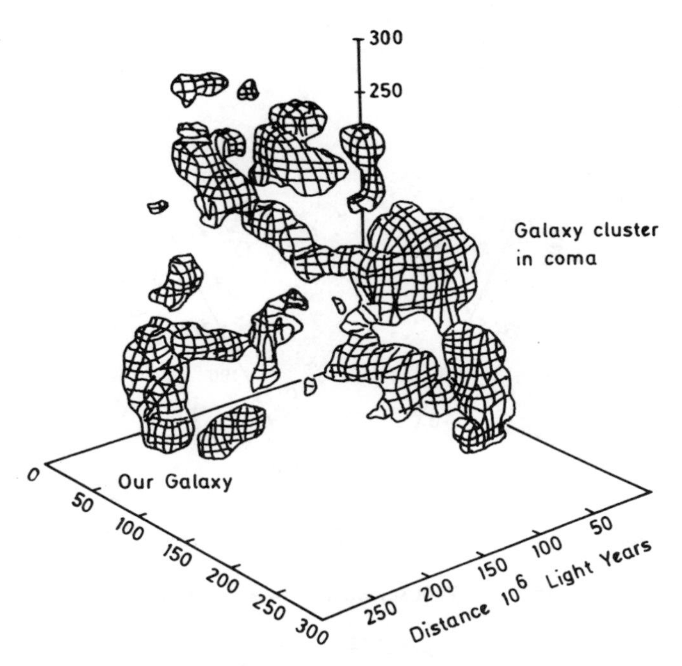

Fig. 1.9. The distribution of galaxies on the scale of superclusters is mesh-like.

collapse, that is, it takes a certain time for the gas cloud to contract into stable objects. This time is called the *contracting time* or the *collapsing time*.

We can therefore give the following explanation that smaller-scale objects have symmetrical forms because they have already completed the contracting process, whereas superclusters do not have symmetrical forms because since they require so long a collapsing time, they have, up to now, not yet completed the collapsing or contracting process. Indeed, as seen in Fig. 1.11, the superclusters still appear to be in the process of contraction.

Let us now make a slight, quantitative estimate. Let us calculate the collapsing time of a gravitating system. Consider a spherical system as shown in Fig. 1.12. Its radius is R and its total mass is M. Under self-gravitation, the system contracts towards its centre. Let P be a typical point on the surface of the sphere. Under the attraction of mass M, its infall velocity v is approximately

$$v \sim \sqrt{\frac{GM}{R}}$$

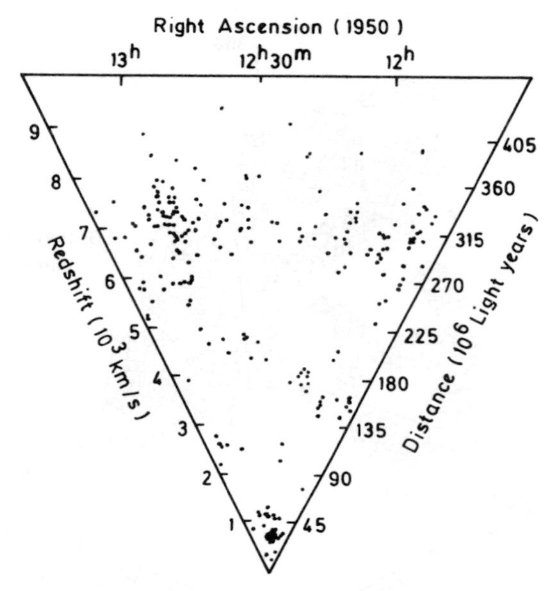

Fig. 1.10. Void surrounded by superclusters.

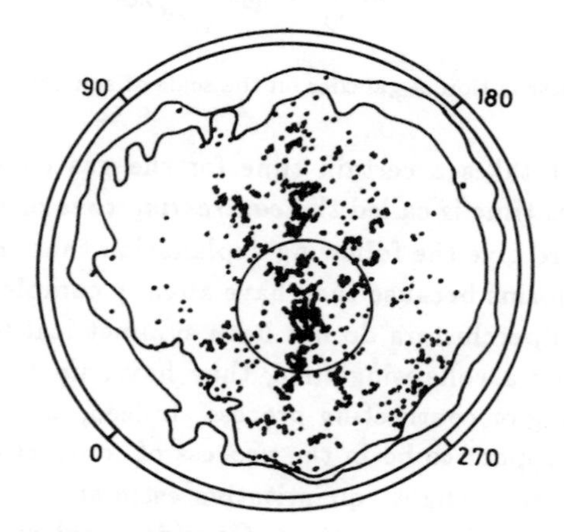

Fig. 1.11. Superclusters still seem to be collapsing.

where G is the gravitational constant. At this velocity, the time taken to fall into the centre is approximately

$$t_c \sim R/v \sim (R^3/GM)^{\frac{1}{2}} \ .$$

Assuming matter to be uniformly distributed inside the sphere, so that the material density within can be expressed as $\rho \sim M/R^3$, the above formula for

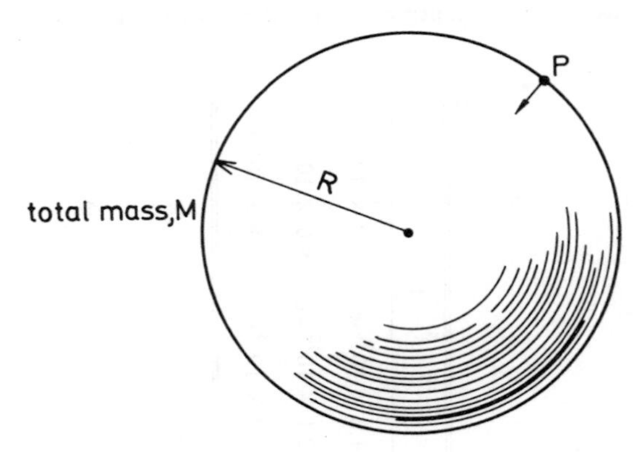

Fig. 1.12. A spherical system with radius R and total mass M is collapsing towards its centre under self-gravitation.

t_c can then be re-written as

$$t_c \sim (G\rho)^{-\frac{1}{2}} .$$

This t_c is called the time-scale of gravitational collapse; it marks the time required for the process of gravitational collapse. Once we know the density of the system, we can calculate its time-scale of collapse.

Table 1.1 lists the collapsing times of several cosmic systems. We see from this that the smaller-scale systems indeed have shorter collapsing times, and the superclusters, the longest. Therefore, if the nebulae forming the superclusters have existed for a time longer than the t_c for galaxies but shorter than the t_c for superclusters, then the lack of symmetry in the latter can be understood.

Finite Lifetime

The above conclusion can also be stated as follows: the galaxies could not have existed for an indefinite time. According to Table 1.1, the age of the gas cloud should be less than 40 billion years, or in other words, 40 billion years ago, no gas clouds existed in the universe.

The universe is made of matter and the matter in today's universe is the same as the matter in the original nebula. Therefore, the finite age of the nebula is an intimation that the matter in the present-day universe also has a finite age.

Next, if all the matter in the universe has only a finite lifetime, then is it any longer meaningful to talk of the universe being infinite in time? For time

Table 1.1. Collapsing times of gravitational systems.

Table 1.1. Collapsing Times of Gravitational Systems

System	Mass	Scale (radius)	Density (g/cm^3)	t_c
Solar System	2×10^{33} g	50 A.U. $\sim 7.5 \times 10^{14}$ cm	4.7×10^{-12}	1.8×10^9 sec $\sim 5.7 \times 10$ yr
Milky Way	$10^{11} M_\odot$	1.25×10^4 pc $\sim 3.9 \times 10^{22}$ cm	3.5×10^{-24}	2.1×10^{15} sec $\sim 6.6 \times 10^7$ yr
Superclusters	$10^{16} M_\odot$	1.3×10^8 l.y. $\sim 1.2 \times 10^{26}$ cm	1.1×10^{-29}	1.2×10^{18} sec $\sim 3.8 \times 10^{10}$ yr

$M_\odot$ = solar mass. l.y. = light-year. pc = parsec.

cannot be separated from matter, and infinitude of time is meaningful only when measured by existing matter! We may seem to have gone too far in the way of *wild* inferences, but, at any rate, they are not entirely groundless.

In summary, research centered on the question of the sky falling has resulted in the following developments in science:

1. The recognition that a universal gravitation is obeyed by heavenly and earthly bodies alike.
2. The recognition that the heavenly bodies are evolving.
3. The recognition that the universe may not exist forever, rather, that it may have a finite life.

The first two points have been generally accepted. The following chapters of this book will show systematically that modern cosmology is providing more and more evidence for the third point. The latest piece of evidence now follows to conclude the present chapter.

Proton Decay

We all know that matter is made of *elementary particles* (or simply *particles*). There are many kinds of elementary particles. The majority of particles will decay into other particles, i.e., they cannot exist indefinitely without change. Particles that exist stably are very few; formerly, it was thought only electrons and protons do so. The matter surrounding us is almost entirely made of electrons and protons. This fact shows that only stable particles have survived to the present while unstable particles have decayed and disappeared in the course of time.*

The newest development states that *protons* may be *unstable.*

In the early 70's, theory already predicted the instability of the proton. It may decay in the following manner:

$$p \longrightarrow e^+ + \pi^0$$
$$p \longrightarrow \pi^+ + \bar{\nu}$$

where p represents a proton; e^+ a positron; π^0 and π^+ are π-mesons with zero and positive electric charges respectively; and $\bar{\nu}$ an antineutrino. Preliminary theoretical estimates for the above decay give a time estimate of not less than 10^{30} years.

---

*Neutrons in the free state are also unstable, with a decay time of about 10 minutes, but may be stable in bound states. Hence, in Nature, there are very few free neutrons, but there are a lot of them inside the atomic nuclei.

Already a number of experimental groups are engaged in measuring the lifetime of the proton. The experiment is not easy to perform, because the proton life is so exceedingly long that decay events are extremely difficult to observe and also because it is not easy to exclude the effect of outside interferences. Hence, up to the present, no sure results have been obtained. Nonetheless, some groups have claimed to have observed instances of proton decay. The proton life given by these observations is roughly in the range of 10^{31}–10^{32} years.

Here, we shall not describe the details of the experiment; we shall only discuss its cosmic implications. If the above experiment is correct, then it is of great value to cosmology. Even though proton life is very long, as long as it is finite, protons cannot exist forever. The matter surrounding us has protons as its main component and if protons cannot exist forever, then the present existing matter in the universe must have been produced no more than a finite time ago. It cannot be infinite in time. This is the same conclusion as that obtained from the problem of the falling sky.

Apparently, not only is the falling sky of value to the men of Qi, the collapsing Earth is likewise no wild fantasy. Since even protons decay, an Earth that is mainly made of protons surely cannot escape collapse, even if it has been able to escape destruction by other agents.

Chapter Two

EXPANSION WITHOUT CENTRE

The Copernican Principle

All religious models of the cosmos have centres. This is so in the West as well as in the East. There is a model of the world made of clay in front of the *Hall of Mathematics* (*Shu Xue Dian*) of the famous Buddhist temple Yong He Gong in Beijing, which shows the Buddhist's view, namely, that the centre of the universe is Mount Xumi.

One of the basic starting-points of modern cosmology is the notion that the universe has no centre, or, more precisely:

> *No points in the universe are preferential; all positions carry the same weight.*

This doctrine is called the *Copernican Principle.* As a matter of fact, it is not found in Copernicus' writings. We named it after him to commemorate his pioneering negation of the geometric view. In his heliocentric proposal, the world model still had a centre. Only later was the Sun found not to be the centre of the world. Though the Sun has a predominant position in the solar system, it is merely an ordinary star when viewed on a larger scale and its position is in no way privileged. Going one step further, we recognize that no point in the universe is preferential. This is the origin of the notion of the Copernican Principle.

Another basis for the principle comes from mechanics. In mechanics, we are used to treating problems in the following way: when studying falling bodies, we consider only the gravitation of the Earth and neglect all other influences; when studying the planets, we consider only the gravitation of the Sun and neglect all others. That is, we study each object in isolation. This procedure is suspect in principle, for all the objects in the universe participate in gravitation and, since gravitation is a long-range force, the effect on a falling body due to all other objects in the universe is not necessarily smaller than that of the Earth. The effect on a planet due to all the other bodies of the universe need not be smaller than that of the Sun. However, the results we obtain when we neglect *the effect of all the bodies of the universe* are very good. This shows that the net total effect of all cosmic bodies on the falling body or the planet is zero or very small.

From the properties of gravitational force, we know that only when the matter in the universe is uniformly distributed, its total combined effect on falling bodies or planets is zero. More generally, the reason why we can treat each system in isolation in mechanics and dispense with the consideration of the effect of the other parts of the universe is precisely because the universe is uniform, with no privileged points.

Observational evidence supports the uniformity of the universe, i.e., there is little anisotropy in the microwave background radiation. We shall discuss this in detail in Chapter 7.

We shall now discuss some deductions from the Copernican Principle.

Geometry of the Round Table Conference

The Copernican Principle requires that all positions have equal rights. The term *equal rights* is political in origin. Its implication is the equality of human beings and the interesting thing is that the equal rights of humans sometimes has to be symbolized by the equal rights of positions.

The story goes that long ago, some kings or queens of equal status were to meet one another. This set their retinues in a flurry – how were they to arrange the seats so as not to show any precedence? Eventually they found a solution: by arranging the seats around a round table this difficulty was satisfactorily overcome. This is the origin of the round table conference.

Figure 2.1 shows five participants in a round table conference. From a geometrical point of view, a round table has the following properties:

1. Viewed as a whole, none of the participants has a privileged position, that is, no one is the centre of the conference.
2. Each participant sees two other participants sitting symmetrically on either side, that is, the seating arrangement seen by each participant is the same.

3. Each participant feels that he/she is at the centre and that the others are surrounding him/her with himself/herself as the centre.

In brief, the subtlety of a round table conference is that it achieves a state of no centre by making everybody think they are the centre. This is the essence of the geometry of the round table conference.

Fig. 2.1. All five participants of a round table conference have equal rights. None of the participants has a privileged position. Each participant sees the same picture with respect to themselves. Namely, either no one is the centre of the conference or everyone perceives themselves as the centre.

The geometry of the universe is very similar to that of the round table conference.

The Cosmic Picture

The geometry of the round table conference is determined by the relations among the participants; likewise, the geometry of the universe is determined by the relations among the stars. Every star is a participant of the universe. Each can carry out a series of observations on the distribution and motions of the surrounding heavenly bodies and obtain a picture of the heavenly bodies in space and time. This picture is called a *cosmic picture*.

As in the round table conference, if all positions have equal rights, then the cosmic pictures seen by all stars should be the same, that is, Property 2 listed above should be obtained from each star. The same distribution and motions of stars are seen.

This is the first deduction from the Copernican Principle. Obviously, Ptolemy's cosmology is at variance with this deduction. In the Ptolemaic system, different observers will see different pictures of the cosmos. For example, from the Earth we will see all the heavenly bodies in regular revolutions while beings on another star will see far more complicated motions of their starry sky. In this system, the Earth is the centre.

The conclusion of the sameness of all cosmic pictures has far-reaching implications. It shows that we can study the entirety of the universe from its local properties. The aim of cosmology is to study the whole of the universe, but our observations are limited to a local region, that is, we can only take the Earth as the starting point of our observations and what we see is the cosmic picture centered on the Earth. Hence, it seems very difficult to study the whole, but the above conclusion tells us that the cosmic picture seen from the Earth is a typical picture. By *typical*, we mean it clearly possesses common properties, which of course, represent in a way global properties. This thus opens up the path to knowledge of the whole universe from local, Earth-bound observations.

An Unremarkable Deduction

Each participant of a round table conference sees the other participants placed at equal intervals on either side or in a uniform distribution. The cosmic situation is similar: on each star, the observer sees other stars uniformly placed around.

We now give a rigorous proof of this point.

In Fig. 2.2 O and O' represent two observers, each observing the density distribution of stars. Observer O obtains the result $\rho(\mathbf{r}, t)$, meaning that at time t, ρ stars are in unit volume at location $\mathbf{r}$ from O. Similarly, O' finds the result $\rho'(\mathbf{r}', t)$, where $\mathbf{r}'$ represents the location $\mathbf{r}'$ from O'.

According to the Copernican Principle, the pictures seen by O and O' should be the same. That is to say, the density at $\mathbf{r}'$ from O should be the same as that at $\mathbf{r}'$ from O', i.e.,

$$\rho'(\mathbf{r}', t) = \rho(\mathbf{r}', t) \ . \qquad (2.1)$$

For a point P in space whose position is $\mathbf{r}$ relative to O and $\mathbf{r}'$ relative to O' (see Fig. 2.2), we obviously have

$$\mathbf{r}' = \mathbf{r} - \mathbf{c} \qquad (2.2)$$

where $\mathbf{c}$ is the position vector of O' relative to O. Since P is a point, the density at this point should be the same whether seen from O or O', i.e.,

$$\rho'(\mathbf{r}', t) = \rho(\mathbf{r}, t) \ . \tag{2.3}$$

Comparing the expressions (2.1) and (2.3), we have

$$\rho(\mathbf{r}', t) = \rho(\mathbf{r}, t) \ . \tag{2.4}$$

Then from (2.2), this becomes

$$\rho(\mathbf{r} - \mathbf{c}, t) = \rho(\mathbf{r}, t) \ . \tag{2.5}$$

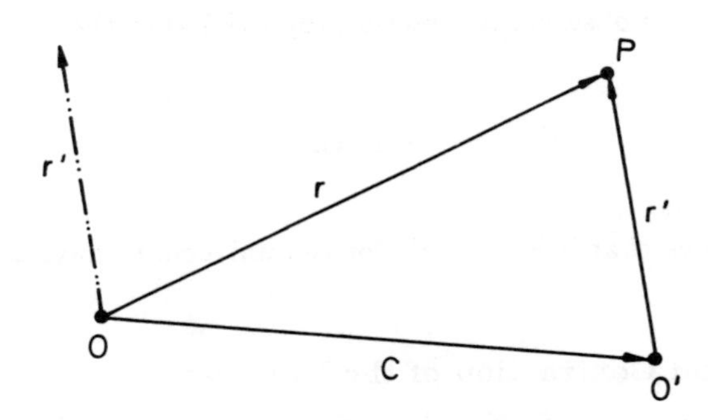

Fig. 2.2. The geometry of two observers O and O′ and the observed point P.

But $\mathbf{c}$ is arbitrary and $\mathbf{r} - \mathbf{c}$ can represent an arbitrary point. Thus, the expression (2.5) says that, seen from O, the density at $\mathbf{r}$ is equal to the density at any point, or that the density is the same everywhere.

It is therefore proven that once positions have equal rights, uniform density necessarily follows. There is a direct logical relation between *equal rights* and *uniformity*. This result is rather unremarkable and it seems that one could pass from *equal rights* to *uniformity* without such an elaborate proof. Actually our real aim is to use this unremarkable deduction as an introduction to the above method of rigorous proof in preparation for deducing remarkable conclusions.

A Variable Round Table

Let us come back to the round table conference. Suppose the table is elastic and its radius can increase or decrease. The positions of the participants then vary accordingly (Fig. 2.3).

Obviously, whether the radius of the round table increases or decreases, the geometrical relations among the participants still keep the three properties mentioned earlier, i.e., the principle of equal rights of positions is still maintained. In other words such variations as expansion or contraction of the round table are compatible with the principle of equal rights.

As the round table expands (contracts), each participant from their own point of view still always sees themselves as the centre, and the people on either side symmetrically recede (approach), i.e., the picture one sees is an expansion (contraction) about oneself as the centre. This expansion (contraction) is very orderly, with participants closer to the observer having small velocities and vice versa. Quantitatively speaking, the velocity $\mathbf{v}$ of a participant relative to the observer is directly proportional to the distance d between them,

$$\mathbf{v} = kd ,$$

with k a constant.

We now prove that it is possible for the universe to have a similar kind of motion.

Expansion and Contraction of the Universe

Refer again to Fig. 2.2. Now both O and O' observe the velocities of the surrounding stars relative to them. Let the result obtained by O be $\mathbf{v}(\mathbf{r}, t)$, or the velocity of the star at location $\mathbf{r}$ relative to O at time t. Similarly, O' results in $\mathbf{v}'(\mathbf{r}', t)$.

Following the argument leading to (2.1), we can likewise argue that the velocity pictures seen by O and O' should be the same, i.e.,

$$\mathbf{v}'(\mathbf{r}', t) = \mathbf{v}(\mathbf{r}', t) . \tag{2.6}$$

Also, for the point P, its velocity relative to O is $\mathbf{v}(\mathbf{r}, t)$, relative to O' it is $\mathbf{v}'(\mathbf{r}', t)$. Additionally, the velocity of O' relative to O is $\mathbf{v}(\mathbf{c}, t)$. According to the classical law of velocity composition*, the two velocities $\mathbf{v}(\mathbf{r}, t)$ and

---

*Regarding the classical law of velocity composition, cf. Fang Li Zhi and Chu Yao Quan, *From Newton's Law to Einstein's Relativity*, Ch. 3, World Scientific & Science Press, 1987.

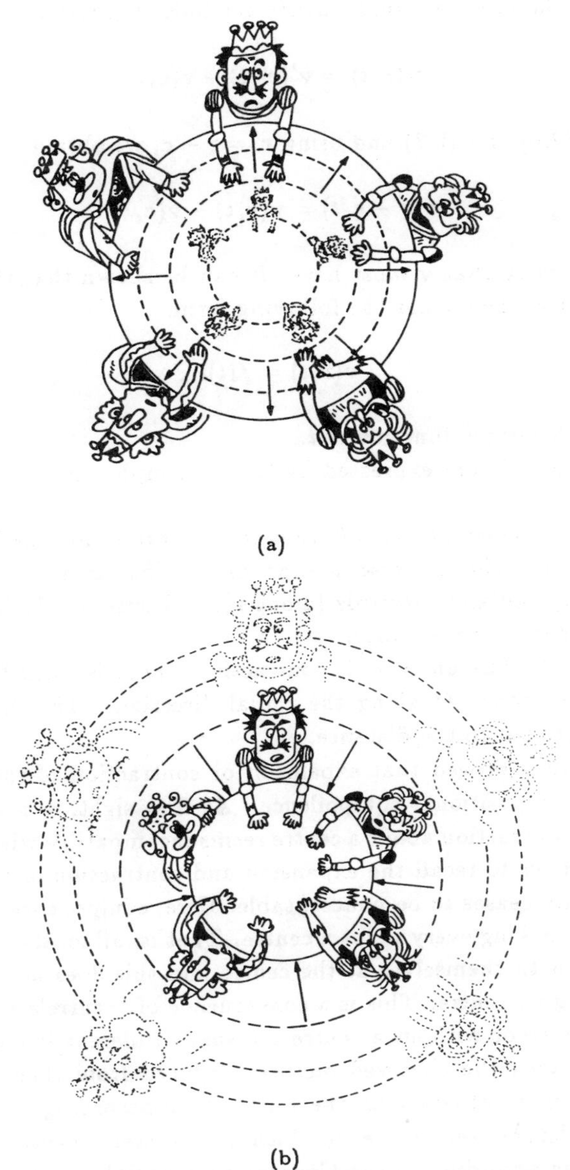

(a)

(b)

Fig. 2.3. An expandable and contractable round table. In (a) contracting and (b) expanding, the relationship among the participants are always of equal rights and all participants will see the same picture of the expansion (or contraction) with respect to themselves.

$\mathbf{v}'(\mathbf{r}', t)$ and the velocity $\mathbf{v}(\mathbf{c}, t)$ satisfy the following relation:

$$\mathbf{v}(\mathbf{r}, t) = \mathbf{v}'(\mathbf{r}', t) + \mathbf{v}(\mathbf{c}, t) . \tag{2.7}$$

Comparing (2.6) and (2.7) and using $\mathbf{r}' = \mathbf{r} - \mathbf{c}$, we obtain

$$\mathbf{v}(\mathbf{r} - \mathbf{c}, t) = \mathbf{v}(\mathbf{r}, t) - \mathbf{v}(\mathbf{c}, t) .$$

This is a property that $\mathbf{v}$ must have. It can be shown that this requirement is satisfied only when $\mathbf{v}$ has the following form:

$$\mathbf{v}(\mathbf{r}, t) = f(t)\mathbf{r} , \tag{2.8}$$

with $f(t)$ an arbitrary function of t.

The velocity picture expressed by (2.8) is simple and can be divided into 3 cases:

1. $f(t) = 0$, hence $\mathbf{v}(\mathbf{r}, t) = 0$, that is, the entire universe is static.
2. $f(t) > 0$. The universe is expanding. That is, seen from O, all the stars fly radially outwards (along $\mathbf{r}$), and with a velocity directly proportional to the distance.
3. $f(t) < 0$. The universe is contracting. That is, seen from O, all the stars approach O along the radial direction. The velocity is again proportional to the distance.

We have thus proved that expansion or contraction of the universe is a logical result of positional equipollence. At first sight, a cosmic picture of expansion or contraction about a centre seems incompatible with equipollence. But we only have to recall the expansion and contraction of the round table and this picture ceases to be unacceptable. Here, equipollence or equal rights is realized by making every point a centre. That is, all local observers see the same picture with themselves at the centre and this does not contradict the totality having no centre. This is a consequence of "centrelessness".

Theories with or without a centre are vastly different in concept, but the difference between the observed pictures deduced from them is very small. You see, if O is an observer on the Earth, then according to the Ptolemaic system, the picture seen is one in which all the stars revolve about the observer, whereas according to the Copernican Principle, it is one in which all the stars fly away from or collapse inwards towards the Earth. The difference between the two is only that vertically upward or downward motion has replaced motion in circles.

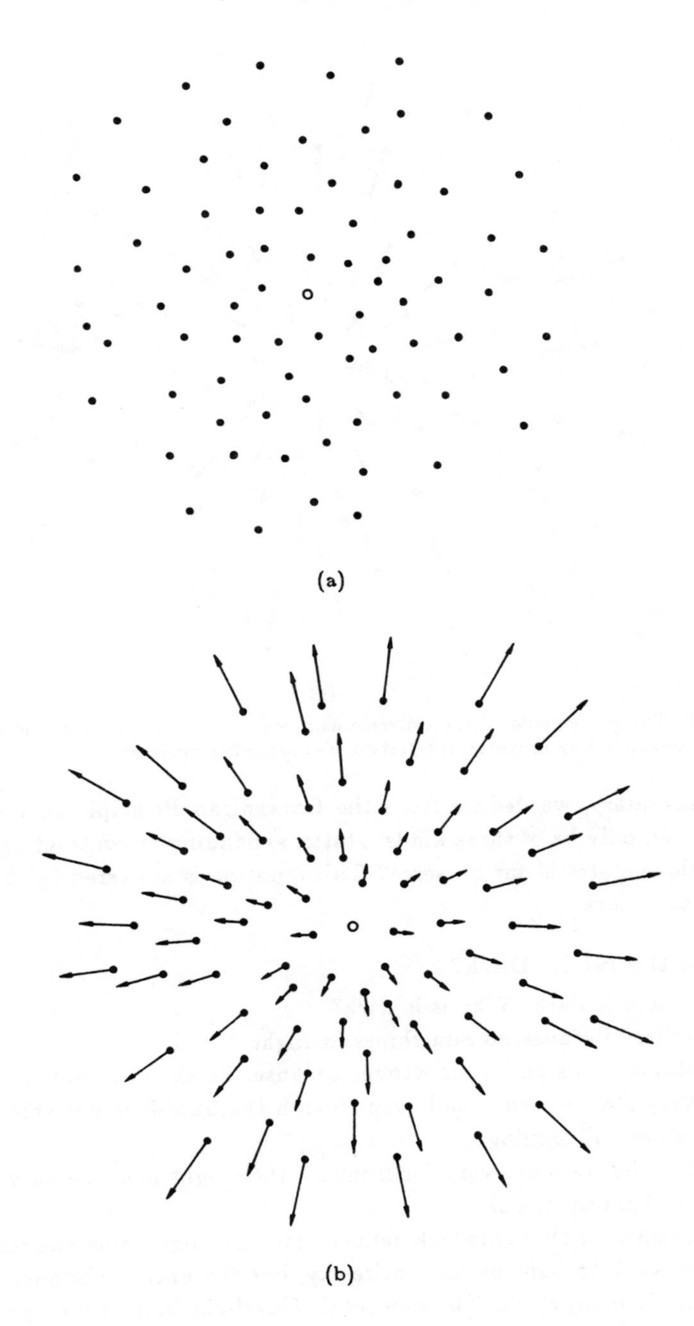
(a)

(b)

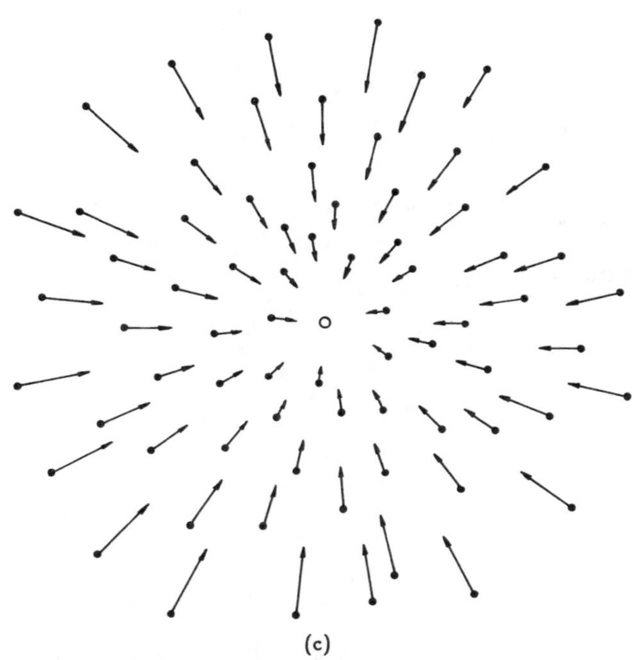

(c)

Fig. 2.4. The movements of the universe as a whole. (a) $f(t)=0$, a static universe; (b) $f(t)>0$, an expanding universe; (c) $f(t)<0$, a contracting universe.

In summary, we deduce from the Copernican Principle that the cosmic picture can only be of three kinds: static, expanding or contracting. In which of the three states is our universe? This question is answered by the fact that *the night is dark.*

Why is the Night Dark?

The night is dark. Why is it dark?

A replies: Because no Sun shines at night.

B retorts: This answer is wrong, because the sky does not have just one sun. Every star is a sun, and even though the Sun does not shine at night, the stars are still shining.

A: No, the stars are very far from us, their light must be very weak, not enough to light up the sky.

B: Again your thoughts lack deliberation. The light from one star is, to be sure, too weak to light up the entire sky, but the universe is huge and there are infinitely many stars. The sum total of starlight from infinitely many stars can well light up the whole sky,

A: I do not think so, but . . .

B: O.K., let me give you a proof.

B's proof now follows.

First consider one single star, (Fig. 2.5(a)). This star emits E photons per unit time so the number of photons per unit time received by unit area at distance r is

$$E/4\pi r^2 \; . \tag{2.9}$$

If the Earth is at distance r from this star and since the Earth has a cross section of πR^2, R being the Earth's radius, the number of photons falling on the Earth per unit time is then

$$(E/4\pi r^2)\pi R^2 = ER^2/4r^2 \; . \tag{2.10}$$

When r is large, this value is very small, showing that the contribution from a single star is minute (Fig. 2.5(b)).

Now consider the entire universe. Suppose space is infinite and contains an infinity of stars, uniformly distributed, that is, suppose there are N stars per unit volume. Let us calculate the number of photons sent to the Earth by this infinity of stars.

With the Earth as the centre, divide the sky into various spherical shells (Fig. 2.5(c)), of thicknesses $d_1, d_2, d_3, \ldots$. Consider a typical shell, say, shell No. 3. Its volume is $4\pi r^2 d_3$ and the total number of stars it contains is

$$N4\pi r^2 d_3$$

Each star in this shell contributes the amount shown in (2.10). Hence, the number of photons sent by the stars of this shell in unit time to the Earth is

$$(ER^2/4r^2)N4\pi r^2 d_3 = \pi R^2 NEd_3 \; .$$

Therefore the total number of photons sent to Earth by all the stars of the universe in unit time is

$$\pi R^2 NE(d_1 + d_2 + d_3 + \ldots) = \infty \; ,$$

so it is infinite! For $d_1 + d_2 + \ldots$ is infinite. This value is far greater than the number of photons sent to Earth by the Sun in unit time. Therefore, the night should be bright.

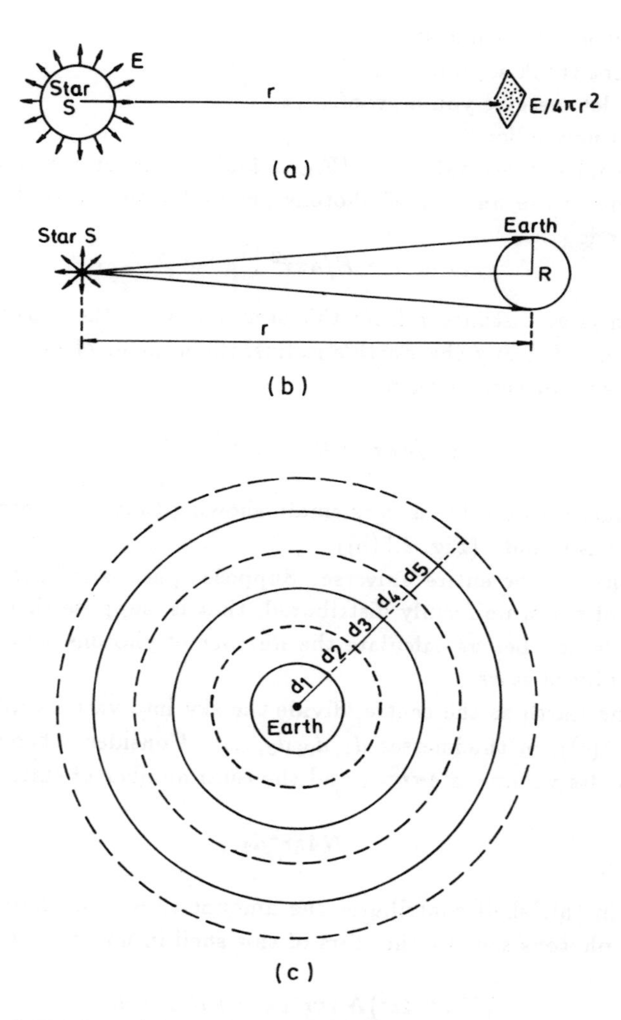

Fig. 2.5. Olbers' paradox. (a) Star S emits E photons per unit time. An observer at distance r from S will receive photons of $E/4\pi r^2$ per unit time and unit area; (b) Photons from star S will be received by the Earth; (c) The sky is divided into a series of spherical shells of thickness d_1, d_2, d_3, ... with their centre on the Earth.

This strange result was noted probably as early as some 170 years ago, and it is called *Paradox of the Dark Night*, or *Olbers' Paradox*. Through the intervening years, a variety of attempts have been made to resolve this paradox but all have failed. It was not until the advent of modern cosmology that a correct answer was found to the question as to why the night is dark.

Before we describe the correct answer, we should point out another significance of this problem. Dark night is a phenomenon seen on Earth. It is a local phenomenon, but the above proof shows that it is closely related to the question of whether the universe contains infinitely many stars. Therefore, it is a very good example of the ability of deducing some properties of the whole universe from local observations.

The World Horizon

The darkness of the night is due to the expansion of the universe. This is the answer given by modern cosmology.

We shall use Newtonian mechanics to prove this result. In fact, throughout this chapter, we shall confine ourselves to Newtonian mechanics and shall not deal with relativity. Hence the photon discussed in this chapter is essentially Newton's corpuscule of light.

Newton's light corpuscule is a classical particle and it obeys Newton's law or the classical law of velocity addition. In Fig. 2.6, O is the observer and O' is a light source which is moving outwards with respect to O with velocity v. O' emits a photon towards O with an emission speed c. The speed measured by O should then be

$$c' = c - v \ .$$

We therefore see that the speed of O' is as large as c or larger, that is, if $v \geq c$, then the photon emitted by O' cannot reach O, for the photon will then no longer move towards O.

The velocity of universal expansion is given by the formula (2.8). It is larger, the greater the distance. According to (2.8), at distances

$$d \geq c/f(t) \ ,$$

all the stars will be moving away from O with speeds greater than the speed of light, hence the photons they emit will never reach O. In other words, observer O can only see the stars within the distance $c/|f(t)|$ and cannot see any stars beyond. This distance is called O's "horizon". O can only see within this horizon and not beyond.

Thus, there is only a finite number of stars shining on the Earth and as long as the number of photons from these stars arriving on the Earth in unit time is less than that from the Sun, night will be darker than day.

If the universe is contracting or static, then there will be no horizon and we shall not be able to explain the darkness of the night.

Fig. 2.6. The law of velocity addition in classical mechanics states that the speed of the light emitted by an airplane at speed **v** will be c–**v** with respect to the observer *O*.

Therefore, we have proven that the dark night is a result of the universe expanding.

It should be stressed that the above proof is based on Newtonian theory. In fact, in problems involving the speed of light, we have to consider relativity. The interesting thing is, after taking relativity into account, we still obtain, it seems, exactly the same results.

Hubble's Discovery

The darkness of the night is the only indirect evidence for universal expansion; direct evidence is provided by the redshift of spectral lines of galaxies discovered by Hubble.

In the 1920s Hubble proved that spiral galaxies are not objects within our own galaxy; rather, they are independent systems of stars very similar to our Milky Way system. Scientists realized that the universe was full of galaxies.

Later, Hubble carried out investigations on the spectral lines of a number of galaxies in our vicinity. Dark lines were found in most of these spectra. They showed that light waves of certain definite wavelengths had been absorbed by the atoms in the atmosphere of the stars. Each chemical element produces a characteristic set of absorption lines, and the wavelengths of these lines have been determined by measurements in laboratories.

When a galaxy moves away from the observer, the wavelength of each spectral line increases. This is the so-called *Doppler effect.* An increase in wavelength means the spectral line is shifted in the direction of longer waves, and as such is called a redshift. Hence, by measuring the *redshift* we can calculate the velocity of recession. If the galaxy moves towards the observer, the wavelengths of spectral lines decrease. The lines shift towards the blue end. This is called *blueshift.*

In 1929, Hubble studied 29 galaxies. Their distances were known. After measuring the spectral lines of these galaxies, he discovered that all these galaxies were moving away from us and, also, that their velocity of recession was directly proportional to their distance, that is

$$v = kd \ . \tag{2.11}$$

This discovery greatly excited Hubble. He wrote:

> *... such scanty material, so limited in its distribution, yet the result is so definite.*

This *definiteness* refers to the result being a direct confirmation of the prediction of universal expansion. In astronomy, it is not easy to make precise predictions; it is even harder to verify such predictions, yet, for the grand expansion of the universe, the prediction was successful both qualitatively and quantitatively.

Since then, the constant k in the formula was called *Hubble's constant.* In Hubble's writings, k was still used to denote this constant. Later, to remember him, it was written as H_0.

The first value of k given by Hubble was later found to be incorrect, but the linear relation between v and d has remained. The measured range has now increased several hundred times since Hubble's time, yet the linear rate of increase is still satisfied by the redshifts of galaxies and their distances. The best value of Hubble's constant known at present is

$$H_0 \approx 150 \text{ km/s per } 10^7 \text{ light-years} \ .$$

That is, galaxies at a distance of 10^7 light-years from us will have a recession velocity of about 150 km/s.

As one astronomer said:

> *The Hubble relation has been recognized as the most outstanding discovery in 20th century astronomy. Like the Copernican revolution 400 years before, it caused another great change in our ideas about*

the universe. It displaced an eternally static universe and confirmed the astonishing fact of universal expansion.

This is a very appropriate summary.

Chapter Three

AGE OF THE UNIVERSE

Eighteen Thousand Years

Western culture does not contain any inkling of universal expansion. Therefore, Hubble's discovery of the redshift phenomenon in the spectral lines of galaxies caused a great stir. Even Einstein regretted that because he could not shake off the influence of tradition, he had narrowly missed the opportunity of discovering the concept of universal expansion himself.

In ancient China, however, there was no lack of Heaven and Earth expanding. For example, in *San Wu Li Qi* (*The Three-Five Calendar*), Xu Zhen wrote:

> *Heaven and Earth are all mixed up like an egg, and Pan Gu is born therein. After eighteen thousand years, Heaven and Earth separate. The clear Yang becomes Heaven and the murky Yin becomes Earth, and Pan Gu is in the middle, undergoing nine changes every day. The spirit is in Heaven, the sage is on Earth. Everyday, Heaven rises by one zhang, and Earth sinks by one zhang. Thus for eighteen thousand years, Heaven becomes exceedingly high, Earth, exceedingly deep and Pan Gu, exceedingly tall. Hence Heaven is ninety thousand li from Earth.*

This vivid theory of universal expansion is not only qualitative, it is also quantitative.

Let us check the numbers. If on each day Heaven rises by one *zhang* and Earth sinks by one *zhang*, then the distance between Heaven and Earth increases by two *zhang* each day, and each year it increases by $2 \times 365 = 730$ *zhang*. According to the ancient Chinese system 1 *li* = 300 *bu* = 150 *zhang*. Hence the yearly increase in the Heaven-Earth separation is $730 \div 150 \approx 5$ *li*. Hence, if 18 000 years have elapsed since the *separation of Heaven and Earth*, the present distance between them should be

$$18\,000 \times 5 = 90\,000 \; li \; .$$

This is precisely the figure given by Xu Zhen, who must have given the matter some careful consideration.

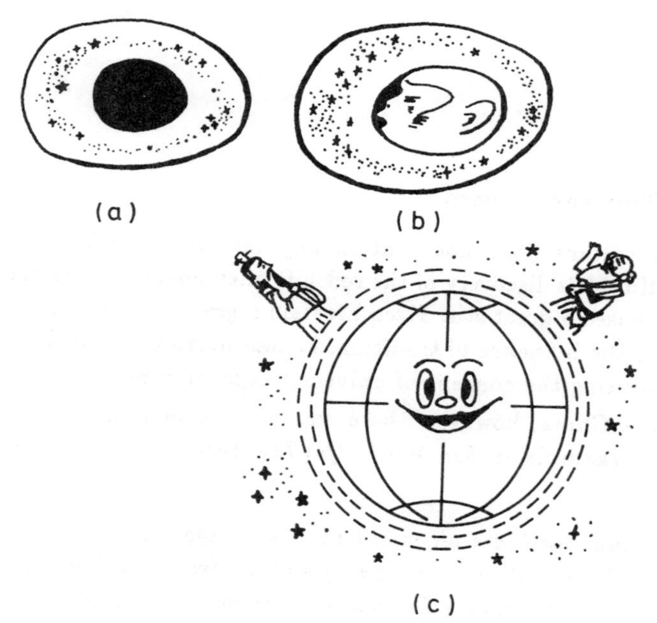

Fig. 3.1. In ancient China, the structure of the universe was considered to be an egg (a), from which the Heaven and the Earth were formed by expansion (b), (c).

This piece of calculation shows that, if the universe is generated through expansion, then the age of the universe is finite. Since *Heaven rises by one zhang everyday*, the *separation of Heaven and Earth* must have taken place *eighteen thousand years* ago.

We can use exactly the same method to calculate the age of the cosmos given by the modern theory of universal expansion. In the last chapter it was stated that galaxies at a distance of 10 million light-years from us have an outward velocity of 150 km/s. We might rephrase this as *Heaven rises by 150 km every second*! Since

$$1 \text{ light-year} = 9.4 \times 10^{12} \text{km}$$
$$1 \text{ year} = 3.16 \times 10^{7} \text{s} ,$$

150 km/s is equal to

$$150 \div (9.4 \times 10^{12}) \times 3.16 \times 10^{7} \approx 5 \times 10^{-4} \text{ light-year/year}$$

i.e. *Heaven rises by 5* $\times 10^{-4}$ *light-years every year.* Thus, the time taken to reach a distance of 10 million light-years is

$$T = (1 \times 10^{7})/(5 \times 10^{-4}) \approx 2.0 \times 10^{10} \text{ years} .$$

This is the counterpart to Xu Zhen's *eighteen thousand years.*

Certainly, some will ask, *if T is the time required to achieve 10 million light-years, would not the time be longer for a greater distance?* There are, however, no grounds for this misgiving, for the law of universal expansion is that the speed of expansion is proportional to the distance. For example, objects 20 million light-years away will be expanding at 300 km/s and so the time taken to reach 20 million light-years is again T.

In short, objects seen today at various distances would have been all piled together about 2.0×10^{10} years ago. At that time, Heaven and Earth had not yet separated, therefore T represents the age of the universe.

If the foregoing argument is correct, an immediately derived conclusion is this: all the objects observed today have a finite age, which should not be greater than 20 billion years (1 billion = 1 thousand million). We proceed to examine this prediction.

Isotopic Chronology

All objects have a finite age.

The objects we observe today are almost all made up of various chemical elements. Hence, if we can just prove that all chemical elements have finite ages not exceeding 20 billion years, then this will strongly support the above prediction.

Chemical elements were, for a long time, regarded as the basic elements of matter, eternal and never changing. Even in 1907 when there was much evidence for radioactive transformation, one of the then most eminent physicists, Lord Kelvin, was still strongly opposed to this idea and denied that elements could evolve. In December 1907 Lord Kelvin died. In January 1908, Rutherford published his theory of radioactive decay. The quick succession of the two events may not have been entirely accidental.

Rutherford's theory is that a radioactive element can undergo changes and become another element. The basic law of this transformation process is this: every radioactive element is reduced by half after a certain, definite time interval. This definite time interval is called its *half-life*. This law of decay is of the same form as the Chinese saying, *a distance of a thousand chi is halved each day*, where the *day* is the half-life. Described mathematically, the law of decay is

$$N = N_0 e^{-\lambda t} \tag{3.1}$$

where N_0 is the amount of the element at time $t = 0$, and N at t. The half-life is then

$$t_{1/2} = \ln 2/\lambda \ . \tag{3.2}$$

Different radioactive elements have different half-lives.

The fact that radioactive elements exist in Nature is itself a proof that some chemical elements have finite ages. For if their ages are infinite, then according to (3.1), they would have long since completely changed into something else and would no longer exist.

Shortly after the discovery of the *law of radioactive decay* Rutherford realized that the evolution of the elements can be used as a kind of *clock* to measure time, and this gradually developed into the isotopic chronology of today. The *radioactive clock* used by archaeologists is ^{14}C. This is because ^{14}C has a half-life of 5570 years, which is happily about the same as the time-scale of human civilization, making it an effective tool for dating various ancient cultural relics.

Cosmology aims to study the entire history of the whole universe, hence it can be called *cosmic archaeology*. It was predicted above that the age of the universe is about 20 billion years. Therefore the clocks to be used in cosmic archaelogy should have half-lives of several billion or several tens of billion years.

One of the radioactive elements with such a long half-life is *mother of radium*.

Measuring the Solar System with Mother of Radium

Mother of radium does not mean the element that generates radium. Already in his early, fundamental paper of 1908 on radioactive decay, Rutherford quantitatively pointed out that "*It seems that the most probable mother of radium is uranium, whose transformation period is of the order of one billion years*".

We now know that there are two kinds of uranium in Nature, ^{238}U and ^{235}U. After a series of transformations, ^{238}U becomes ^{206}Pb. Its half-life is 6.5 billion years, that is,

$$\lambda_{238} = \ln 2/6.5 \times 10^9 \ \ \mathrm{yr}^{-1} \ . \tag{3.3}$$

^{235}U eventually produces ^{207}Pb, with a half-life of 1.0 billion years, or

$$\lambda_{235} = \ln 2/1.0 \times 10^9 \ \ \mathrm{yr}^{-1} \ . \tag{3.4}$$

Using the uranium clock, many rocks on Earth have been measured, as well as lunar rocks and meteorites. All gave the result of 4.55 billion years, which is the age of the formation of the solar system.

Now follows a simple description of how this figure of 4.55 billion years is derived. Readers not interested in this can go straight to the next section.

If, for a certain part of a rock sample, the initial contents of the two kinds of lead and the two kinds of uranium are $(^{206}Pb)_0$, $(^{207}Pb)_0$, $(^{238}U)_0$ and $(^{235}U)_0$, then their respective present contents are

$$\begin{aligned} ^{206}\mathrm{Pb} &= (^{206}\mathrm{Pb})_0 + (^{238}\mathrm{U})_0 - {}^{238}\mathrm{U} \ , \\ ^{207}\mathrm{Pb} &= (^{207}\mathrm{Pb})_0 + (^{235}\mathrm{U})_0 - {}^{235}\mathrm{U} \ , \\ ^{235}\mathrm{U} &= (^{235}\mathrm{U})_0 e^{-\lambda_{235} t} \ , \\ ^{238}\mathrm{U} &= (^{238}\mathrm{U})_0 e^{-\lambda_{238} t} \ . \end{aligned}$$

From these equations we can easily deduce

$$\begin{aligned} ^{206}\mathrm{Pb} &= (^{206}\mathrm{Pb})_0 + {}^{238}\mathrm{U}(e^{\lambda_{238} t} - 1) \ , \\ ^{207}\mathrm{Pb} &= (^{207}\mathrm{Pb})_0 + {}^{235}\mathrm{U}(e^{\lambda_{235} t} - 1) \ , \end{aligned}$$

hence

$$^{206}\mathrm{Pb} - (^{206}\mathrm{Pb})_0 = \frac{^{238}\mathrm{U}(e^{\lambda_{238} t} - 1)}{^{235}\mathrm{U}(e^{\lambda_{235} t} - 1)} \cdot [^{207}\mathrm{Pb} - (^{207}\mathrm{Pb})_0] \tag{3.5}$$

It has been noted that in the solar system, the ratio $^{235}U/^{238}U$ is the same in all rocks, the value being

$$^{235}U/^{238}U = 1/140 . \tag{3.6}$$

In the formula (3.5), all the quantities apart from ^{206}Pb and ^{207}Pb are constants, so the formula expresses a linear relation between ^{206}Pb and ^{207}Pb. From the gradient of the line we can then find t, the time of formation of the rocks. Fig. 3.2 gives an example. It is a plot of $^{206}Pb/^{204}Pb$ and $^{207}Pb/^{204}Pb$ for a meteorite named Allende. Since ^{204}Pb is a stable isotope, a relation between these two ratios is equivalent to a relation between ^{206}Pb and ^{207}Pb. The figure shows that this is a good linear relation. The gradient of the straight line gives the age of the Allende meteorite as

$$T_S = 4.553 \pm 0.004 \, \text{billion years} .$$

Though this figure, 4.55 billion years, is the age of various rocks of the solar system, it is still not the age of uranium itself. Obviously, the age of uranium is greater than the age of the solar system. Since the solar system itself has no mechanism for producing uranium, the uranium of the solar system must have been formed before the solar system.

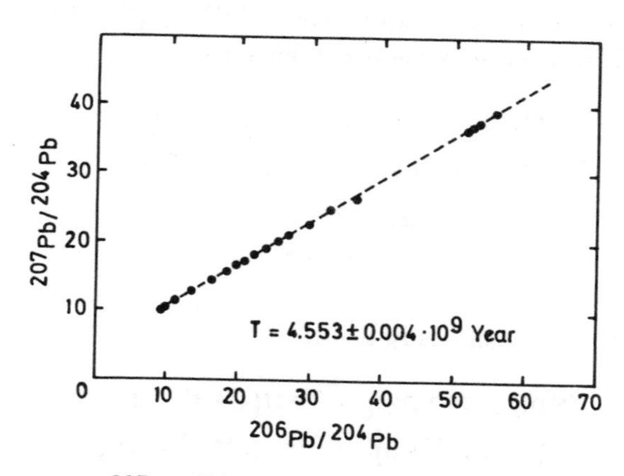

Fig. 3.2. The ratios of $^{207}Pb/^{204}Pb$ and $^{206}Pb/^{204}Pb$ obtained from various inclusions of the Allende meteorite. All measured points lie on a straight line, speaking for a common origin about 4.533×10^9 years ago.

The Origin of Radioactive Elements

Uranium was created in the process of a supernova explosion.

Supernova explosion means an extremely violent explosive process which occurs when a star has evolved into the late stage. In such an explosion, the brightness of the star can increase several hundred thousand times in a few days; that is, the brightness of a single star may become comparable to the brightness of an entire galaxy. Such events are, of course, extremely rare, and in the old Chinese texts covering two thousand years, only a few records can be shown to be possible supernovae. One of the brightest appeared in the year 1006. According to records, one could read in its light.

In the book *From Newton's Laws to Einstein's Relativity*, we also referred to the phenomenon of supernova explosion, where we emphasized that their characteristic features are a proof of the invariance of the speed of light. Here, we shall discuss the consequences in nuclear physics of a supernova explosion.

In such an explosion, a large amount of neutrons are emitted from inside the supernova. The neutrons are easily captured by the matter in the outer layers. Through this process of neutron capture, lighter nuclei are changed into heavier nuclei. Some of the heaviest radioactive elements, such as ^{232}Th ^{235}U, ^{238}U, and ^{244}Pu are all produced this way.

From the properties of neutron capture, we can calculate the relative rate of production of the various heavy elements. For example, the ratio between the two isotopes of uranium is

$$(^{235}\mathrm{U}/^{238}\mathrm{U})_0 = 1.24 \,. \tag{3.7}$$

Let the present ratio observed on Earth be K.

$$^{235}\mathrm{U}/^{238}\mathrm{U} = K \,.$$

Using the formulae (3.1) and (3.7), we see that the following relation should exist between these two values:

$$^{235}\mathrm{U}/^{238}\mathrm{U} = (^{235}\mathrm{U}/^{238}\mathrm{U})_0 e^{-(\lambda_{235}-\lambda_{238})t} \,.$$

This immediately gives the time of generation of uranium as

$$t \approx [\mathrm{In}(1.24/\mathrm{K})]/(\lambda_{235} - \lambda_{238}) \simeq 8.8 \times 10^9 \text{ yr} \,.$$

Similar methods have been used to estimate the time of formation of other radioactive elements and the results show that, indeed, their ages do not exceed 20 billion years.

That radioactive elements must have finite ages is very natural and easily accepted. If the age of the universe is finite, then not only radioactive elements but also all stable elements must have finite ages.

The Age of Stable Elements

In recent times, although the notion of a finite origin of chemical elements has met no such authoritative opposition as that of Kelvin, public recognition of this viewpoint and its related theory has been slow in coming, in a way dissimilar to the fate of the theory of radioactivity. It was not until 1983 when the American physicist W. A. Fowler was awarded the Nobel Prize in physics, that the idea of a finite beginning of the elements won world recognition. For this belated encouragement, the astrophysical world was, of course, very glad ... however, not without a sense of regret. For some of the pioneers of the theory had by then died and so did not share an honor which was partly theirs.

The initial aim of the theory of the origin of chemical elements was to understand the abundance of various elements. By abundance, we mean the relative content found in Nature. For example, iron is plentiful and gold is scarce on Earth. In terms of mass, we say the former has a larger, and the latter a smaller abundance. Fig. 3.3 shows the results of abundance measurements. The horizontal axis is the mass number of the atomic nucleus, and the vertical axis is the percentage by mass of the particular nucleus.

There are many light elements and few heavy elements, but the variation is not completely monotonic.

How did this distribution with many lightweights and few heavies come about? Why is hydrogen so plentiful and noble metals so scarce? Are these abundances eternal? The theory of generation of chemical elements says that the abundances are not eternal, and that there are many lightweights and few heavies precisely because the age of the universe is not that long. This kind of theory of genesis takes as a basic tenet that all heavy elements have been synthesized from light elements. From the point of view of nucleosynthesis, the age of the universe is too short; hence, not many heavy elements have been synthesized from the lighter elements.

Obviously, according to this view of element generation, we can use the abundance measurements of the elements to estimate the times of production of all chemical elements. Thus, all elements, whether radioactive or stable, have finite lives.

What evidence is there to support the view that elements are generated? We may say that all the results of research on stellar evolution are evidence.

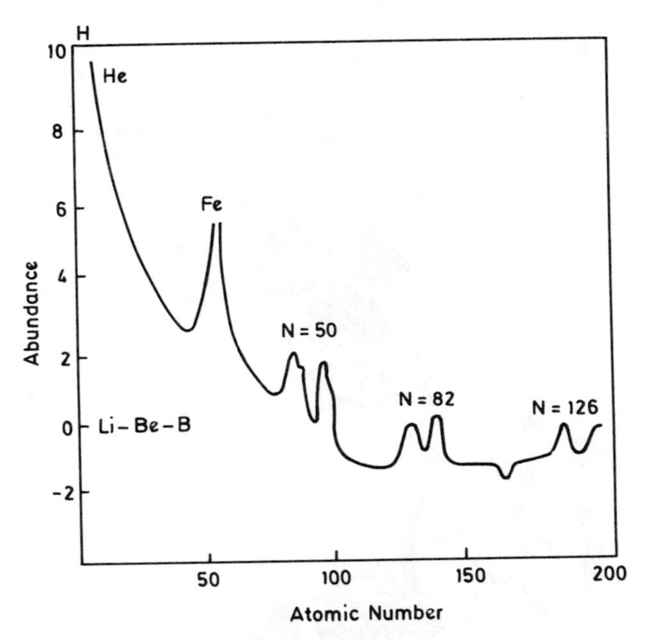

Fig. 3.3. Abundances of various elements.

In order not to stray too far from the main topic of this book, we shall not be giving a systematic introduction to our knowledge on stellar evolution. We shall simply point out that, from the standpoint of element generation, the various types of stars are nothing other than *furnaces* that change light elements into heavy elements. For example, the Sun is continually turning hydrogen into helium, helium stars are synthesizing helium successively into carbon, magnesium, silicon, and so on, up to iron, while red giants and supernovae proceed further to manufacture the various elements heavier than iron. Nearly one hundred elements of the periodic table that exist in Nature are all manufactured in this manner.

During the hundreds of years of the Middle Ages, innumerable alchemists spent their whole lives in their search of *furnaces* that would transform the elements. They failed. Never would they have thought, as they raised their eyes in despair, that every star they saw was just such a furnace!

Two "Small" Pieces of Evidence

There are two *small* pieces of evidence to show that the stars in the sky are indeed the *furnaces* as understood above. One is the resonance reaction

Fig. 3.4. Never would the ancient alchemists have thought that all stars are the furnaces they were looking for.

between ^{8}Be and ^{4}He, the other is the discovery of Tc.

If elements are formed by the lighter ones synthesizing into heavier ones, then the order of their formation will be the same as their order by mass. That is, the elements $A = 1, A = 2, \ldots$ were successively generated. However, laboratory studies of nuclear physics tell us that stable isotopes of $A = 8$ do not exist. For example, ^{8}Be is unstable, having only a very short half-life, hence very little ^{8}Be is present in stars. Now, if $A = 8$ does not exist, how could $A = 9, A = 10$, possibly be formed? This problem is known as the gap at $A = 8$, meaning it is a gap that cannot be crossed and the element synthesis comes to a halt here.

In order not to cut short the synthesis theory, English astrophysicist F. Hoyle conjectured that the reaction $^8\text{Be} + {}^4\text{He} \rightarrow {}^{12}\text{C}$ must be very fast, otherwise it is a resonance reaction. If so, even though ^{8}Be can exist only for a very short time, enough ^{12}C can still be synthesized to jump over the gap. It should be emphasized that this conjecture was entirely inspired by the theory of element synthesis. From the abundances of the elements we can even give a quantitative estimate of the resonance reaction. Subsequently, in the laboratory, this prediction was directly demonstrated to be correct.

Technetium is element No. 43. It is often marked in red in the periodic table to show that it is unstable. Natural technetium is not found on Earth. It's half-life is about 3 million years, far less than the age of the Earth (4.55 billion years). Hence the technetium present at the formation of the Earth has long since disappeared through decay. Moreover, bodies like the Earth cannot possibly produce technetium, hence it is very natural that no technetium is found on Earth. According to what was stated above, elements heavier than iron (atomic number 26) can only be formed in red giants or supernovae, and it is only on such bodies that natural technetium can be found. According to research on stellar spectra, emission lines of technetium have indeed been seen on a certain red giant!

These two *small* pieces of evidence once again support the following "*large*" point of view that you can possibly see the entire universe from local phenomena. Why is Element No. 43 marked in red in the periodic table? Why is there no stable isotope of $A = 8$? These questions are all related to the finite age of the universe.

The Hertzsprung-Russell Diagram of Globular Clusters

Let us return to our topic. Just how great is the age of the stable elements?

Since all elements are produced in stars, the age of the oldest stars is identical to the age of the elements. The best method of measuring the age of stars is by means of the Hertzsprung-Russell diagram (the H-R diagram) of globular clusters.

Thousands, tens of thousands, even hundreds of thousands of stars concentrate together to form a globular cluster. These objects have a spherically symmetric shape or nearly so, hence they are dubbed globular. There are about 500 globular clusters surrounding the Milky Way galaxy. The brightest is known as omega Centauri (Fig. 3.5).

For each star of a globular cluster, we can measure two parameters: its *luminosity* and its *surface temperature*. With the luminosity and surface temperature as the vertical and horizontal axes respectively, we can construct a luminosity – temperature diagram for each globular cluster. This is called the *Hertzsprung-Russell diagram*, in honor of Danish astronomer Hertzsprung and American astronomer Russell who first plotted such diagrams. Figure 3.6 is a typical H-R diagram.

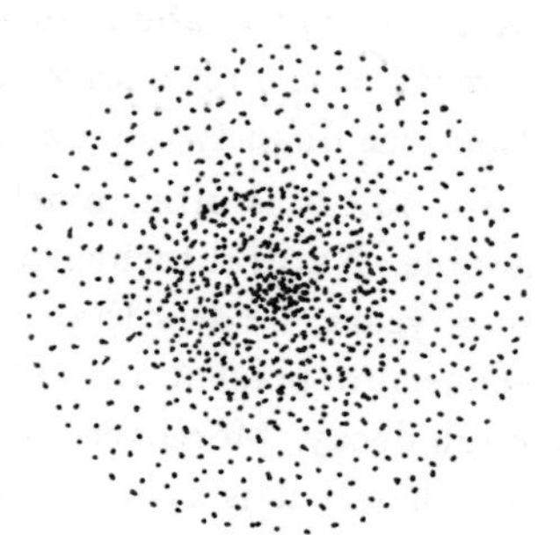

Fig. 3.5. A globular cluster.

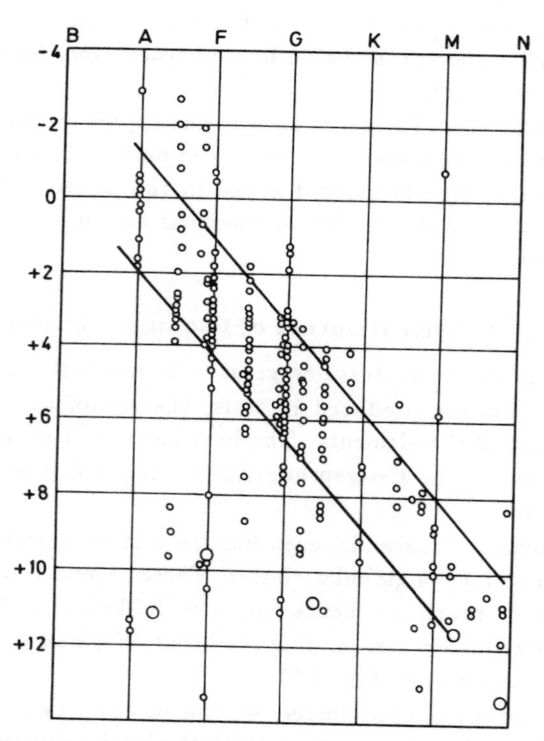

Fig. 3.6. The first Hertzsprung-Russell diagram given by H. N. Russell in 1913, where the vertical axis denotes absolute magnitude and the horizontal one is the surface temperature of stars. The values of the temperature are higher on the left and lower on the right.

Figure 3.6 shows that the luminosity-temperature distribution of stars is very regular. Most of the stars lie on a diagonal line; they are called *main sequence* stars. Their characteristic is that greater luminosities are associated with higher temperatures. A few stars are also found in the upper right hand portion of the diagram. These stars have large luminosities, but their temperatures are not very high; they are what was referred to before as *red giants*. Each H-R diagram of globular clusters shows a clear turn-off point where the main sequence belt turns into the red region.

It was further discovered that for different globular clusters, the turn-off points are different. Figure 3.7 shows a series of typical cases. According to the theory of stellar evolution, different shapes of the H-R diagram mean different evolutionary stages. Generally speaking, the more red giants a globular cluster contains, the greater is its age. Therefore, the more the turn-off point is to the lower right, the greater is the age of the cluster. According to this criterion, we see that, in Fig. 3.7, cluster a has the shortest age, b, c the next shortest, and d the longest.

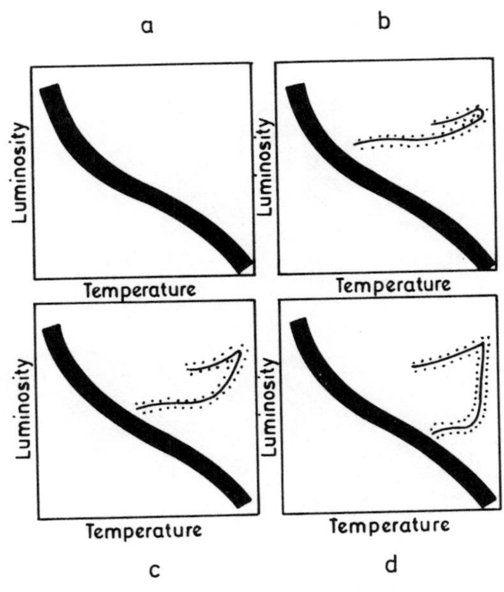

Fig. 3.7. The Hertzsprung-Russell diagram of globular clusters. Different globular clusters possess a different turn-off point. The order of the ages of these globular clusters is a, b, c, d, a being the youngest cluster and d being the oldest.

From the position of the turn-off point we can determine the age of the globular cluster. The age of the oldest cluster determined this way is

$$t_{GC} \sim 1.7 \times 10^{10} \text{ yr} .$$

Again, this is less than 20 billion years.

At this point, we have *nearly* completed the proof of the following statement: all the chemical elements that make up everything were produced within the last 20 billion years. We say nearly completed and not completed, because there are still two reservations:

1. The starting-point of nucleosynthesis is hydrogen and we have not proven that hydrogen was produced within the last 20 billion years.
2. Nucleosynthesis in stars is unable to explain the abundance of some of the lightest elements (helium and deuterium).

Although these two points remain to be discussed, we may say that the theory of generation of chemical elements strongly supports the age predicted by the expansion of the universe.

Chapter Four

FINITENESS AND INFINITENESS OF SPACE

A Question for Idiots

The contents of the two foregoing chapters can be summarized as follows: the universe is expanding; this expansion shows that the age of the universe is finite, that is, the universe is not infinite in time. The question naturally arises, is the universe infinite in space? What answer to this question does the universal expansion give? This is the topic of this chapter.

The question of whether the universe is finite or infinite is not new. Discussion on finiteness and infiniteness can be traced back to the early stage of human civilization. It has been said that, throughout human history, there are two questions on which debate has never ceased: whether abortion can be regarded as murder, and whether the universe is finite. Indeed, in almost all civilizations in the world throughout history we can find, to varying extents, pronouncements on the question of whether the universe is finite or infinite. As we look back through several thousand years of history, we find an almost equal number of scholars believing in a finite or an infinite universe. There seems to have been a continual fluctuation between the two opposing views, lasting almost equally long, right to the present time. Figure 4.1 is a cartoon of the historical situation.

This sketch may give one the impression that this question does not have a believable answer. When Einstein was establishing modern cosmology, he said jokingly:

> *Is the universe infinitely extended? Or is it finite and closed? An answer was given by Heine in one of his poems — only an idiot can expect an answer.*

He wrote this in a letter dated 12 March 1917 to de Sitter. In that year, Einstein and de Sitter were endeavouring to use general relativity to build a model universe. Although it was an unsettling time during the First World War, they corresponded frequently, seriously searching for an answer that only idiots expect. Like the story of the men of Qi worrying about the sky, this demonstrates once again that many of the problems that make physicists and astronomers neglect their food and sleep are viewed, even by the most imaginative poets, as so absurd and wayward that only idiots would waste their lives on such questions.

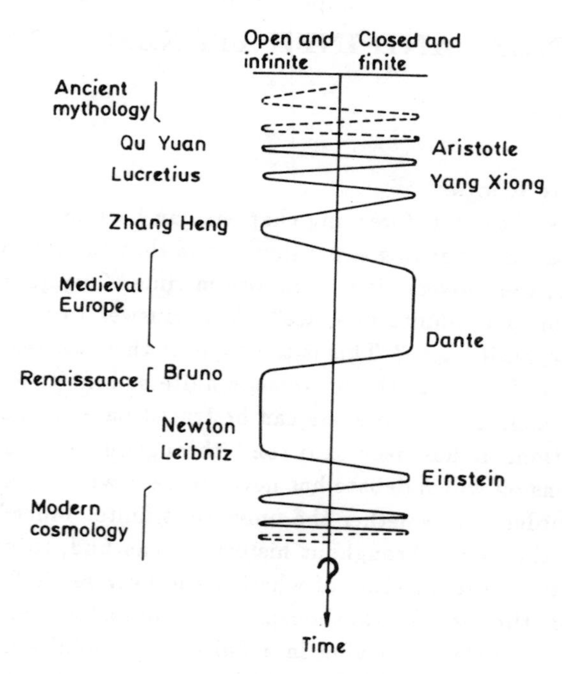

Fig. 4.1. 'Is the universe finite or infinite?' Famous people have held opposing views throughout history.

Heine's answer is superficial and too cut-and-dry, although, as of now, the question of whether the universe is finite or infinite has still not been solved. We have, however, in searching for an answer, acquired deeper knowledge. Let us now take a look back.

The Finite-Infinite Rationalistics

Whether in ancient China or in ancient Greece, almost all the early models of the structure of the universe assumed the universe to be finite and bounded.

The cosmology of Eudoxus and Aristotle is representative of ancient Greece. In this scheme, Earth is a ball that is surrounded by a series of concentric, transparent spheres. The outermost spherical layer is the *sphere of fixed stars.* The entire material cosmos is contained in the sphere of fixed stars.

The Hun Tian theory, developed by Zhang Heng at almost the same time states:

> *The cosmos is like a hen's egg. The body of Heaven is round like a pellet. Earth is like the yolk in the egg residing alone within the Heaven. Heaven is large and Earth is small. There is water on the surface of and inside Heaven. Heaven wraps round Earth like the shell does the yolk. Heaven and Earth both stand on qi and float on water. The circumference of Heaven (measures) three hundred sixty-five and one-quarter degrees. Divided in the middle, one half (measuring) one hundred eighty-two and five-eights degrees comes above Earth, one half goes below, hence half of the twenty-eight lunar mansions are seen and half, hidden. Its two extremities are called the North and South Poles. The North Pole is the middle of Heaven. It is due exact North, and is thirty-six degrees above the horizon, hence a seventy-two degree region across the North Pole is always seen and never hidden. The South Pole is the middle of Heaven and Earth. It is due exact South and is thirty-six degrees below the horizon. A southern region, seventy-two degrees across, is always hidden and never seen. The two poles are over one hundred eighty-two and a half degrees apart. Heaven turns as though driven by a hub, revolving endlessly. Its shape is turbid. Hence the name Hun Tian.*

If we note that ancient Chinese astronomers divided the circumference of the circle into 365.25 degrees and that the capital of the Eastern Han, Luoyang, is located at latitude 35°N, then we see that the above description is the motion of the finite celestial sphere as seen by Zhang Heng himself.

Why were there no infinite models? First, infinite models do not have such direct empirical bases as finite models. Next, it is more difficult for an infinite model to explain the phenomenon of the stars rising in the East and setting in the West.

The Xuan Ye theory of ancient China is one that maintains an infinite universe. Its basic contents are:

> *Heaven is without substance ..., high, distant and limitless, ..., the Sun, the Moon and the stars naturally float and live on the void ... therefore the seven luminaries may pass away or stay, may move either in the direct or the retrograde sense, their disappearances and appearances are not regular, their advances and retreats are different; because they are not tied to anything, their behaviours are all different.*

This school opposes a solid celestial sphere and believes the stars to be floating and living in a void. These are all fine insights. Its view of a *high, distant and limitless* space is particularly original. However, this view never developed into a theory capable of giving a quantitative fit to the rotation of the celestial sphere and the direct and retrograde motions of the planets; hence it amounts, at most, to an idea of infinite space and not a model of an infinite universe.

That the infinite view had difficulty in making quantitative statements did not mean that the finite view had a dominant position in history. In fact the finite universe was subject to constant criticism. One type of rational criticism runs as follows. Finiteness means there is a boundary, and a boundary implies existence beyond. Hence we come to a quandary. Does the boundary have a *beyond*? If it does not, then this contradicts the notion of a boundary; if it does, then this contradicts the idea of the universe itself. For universe means everything: all and sundry, and there cannot be anything outside.

Both Aristotle and Zhang Heng seemed to have noticed this fatal criticism and, interestingly, took very similar steps to overcome this difficulty. Aristotle says that the sphere of fixed stars is no ordinary boundary, that although it has an inside and an outside, the *inside* is the physical space while the *outside* is the world of gods. Beyond the sphere of fixed stars are three more heavenly layers: the crystal, the highest and the pure fire spheres. These are spiritual and soul-like, and so do not occupy physical space. Similarly, Zhang Heng also maintains that the celestial sphere has a beyond; however, "*Of those who crossed this (boundary) and went forth, there has been none known*". His intention is also to place a non-physical existence beyond the boundary of a finite universe, thus resolving the contradiction.

Of course, from the modern scientific viewpoint, such entities as *gods* or *of those there has been none known* cannot serve as sufficient reason in science. Consequently, after the rise of modern science, this viewpoint was rejected.

Rationalistics Continued

Modern science is epitomized by Newtonian mechanics. Newton himself maintained that the universe is infinite. One of the basic starting points of Newtonian mechanics is the existence of absolute space, that is, the infinite Euclidean space. Aristotle's crystal spheres were smashed.

However the debate did not stop.

Newton's contemporary Leibniz also regarded space as being infinite. But he differed sharply from Newton regarding the distribution of stars. Newton believed that stars must be distributed in finite space, the reason being that if they are present in infinite space, they must be infinite in number, and infinitely many stars will have infinite gravitation, thereby making the entire system unstable. Leibniz maintained that stars must be uniformly distributed throughout an infinite space and his reason was that if the star distribution is finite, then the whole physical universe is still bounded and has a centre, and this is unacceptable for any post-Copernican cosmology.

Although Newton and Leibniz are both master pioneers of modern science, the above debate shows that when they studied the finite-infinite question, they had not shaken off the mould of pure, rationalistic thinking since both used the method of proof by the negative.

Kant brought this rationalistic debate to a close. He thought he had found an answer that would settle the question forever — his antinomy of space. Kant says, if we insist 1) that the system of stars is stable, 2) that the system of stars has no centre and 3) that space is the infinite Euclidean space, then we cannot possibly find a logically self-consistent answer. This is to say, not only is it impossible for us to construct a finite model universe without internal contradiction, it is also impossible for us to construct an infinite model universe. The conclusion can only be that the universe can be neither finite or infinite. Hence the finite-infinite question itself has no meaning and we should not discuss it at all.

Slightly preceding Kant in time, Yang Shen of China seemed to have also felt that this debate could not be resolved. He wrote:

> *If heaven has a limit, what lies beyond that limit? Heaven cannot be limitless either because anything having form must have limit.*

He thought both views were apparently reasonable, hence neither was unreasonable.

The analyses by Kant and Yang Shen are rather penetrating. Their theories may be said to have brought to a close the finite-infinite rationalistic debate that had lasted some two thousand years. Their conclusion is that the finite-infinite question itself is impossible.

However, famous for his rigour of argument, Kant's insolvable argument is not all that rigorous. A careful examination will reveal that his argument implicitly used some unproved theses. They are:

Finiteness must mean having a boundary.

Having a boundary must mean finiteness.

Infiniteness must mean having no boundary.

Having no boundary must mean infiniteness.

Kant thought these to be the most ordinary common sense thesis that required no careful discussion.

However, science is often the demonstration of constructiveness in what is regarded as error by common sense.

Unbounded Finiteness

The above statements based on common sense are not correct. In fact, having a boundary does not necessarily mean finiteness. That is, we can have a finiteness without a boundary, as well as an infiniteness with a boundary.

As early as the 5th century B.C., some scholars already discussed whether the Earth is infinite or not. Those of the infinite persuasion had this argument: if the Earth is finite, then surely people would fall off after reaching the boundary? The other school believed, however, that the Earth is finite but has no boundary. Spheres have this property. The surface of a sphere is finite, but it has no boundary. The idea that the Earth is a sphere was born in this period of time.

The Chinese classic Zhuang Zi (Chuang Tzui) contains the following statements attributed to Hui Shi: *The South is infinite, yet is finite*, and *I know the middle of the world is north of Yan and south of Yue*. According to Chinese geography, Yan is in the north, and Yue is in the south. Hence, to regard the *middle* to be *north of Yan and south of Yue* seems to be absurd. However, if we note his "*infinite yet finite*", we shall understand that maybe he was envisaging precisely a finite but boundless Earth.

From the Earth being finite and boundless to the universe being finite and boundless, we need only a very slight generalization in geometry, namely a generalization from two to three dimensions, from finite and boundless two dimensions to finite and boundless three dimensions. This tiny step has taken mankind about two thousand years to accomplish.

In his lecture *On the Hypotheses of the Basis of Geometry* of 1854, Riemann first pointed out that boundlessness of space does not imply infiniteness of space. He says:

> *In our knowledge of the external world, space is assumed to be a boundless three-dimensional manifold. The range of our actual consciousness is being replenished constantly by this assumption and the possible positions of objects we seek are constantly being made determinate by this assumption; with applications in such matters, this assumption is continually being confirmed. It is because of this circumstance that the boundlessness of space has a greater degree of certainty than any other external experience. But we certainly must not infer infiniteness of space from this. On the contrary, if we suppose that the existence of matter is independent of position — and hence can endow space with a constant curvature, then, as long as this curvature has a positive value, however small, space can only be finite.*

These statements by Riemann are entirely analogous to the argument showing that the Earth is a finite and boundless curved surface. To recognize boundlessness of the Earth does not justify the inference of infiniteness; on the contrary, since the Earth has more or less the same positive curvature everywhere, the area of the Earth can only be finite.

Riemann's research has released us from the impasse of Kant's space antinomy and has showed that the finite-infinite question is not impossible. A greater significance of Riemann's theory is that it terminated the era of studying the question of finite or infinite space by pure thinking and started the era of studying this question by the method of verification. For according to Riemann's theory, the universe being finite or infinite is determined by the curvature of space, and the latter is, in principle, a measurable quantity.

In fact, some time before Riemann, Gauss had already discovered certain parts of the Riemannian geometry, but he never dared to publish his results, because he felt that such a curved space geometry ran too much against the grain of common sense and would probably be regarded as a heterodox. However, Gauss was, after all, a scientist who knew that *being against common sense* cannot be a sufficient reason for rejecting a theory. Affirmation or negation in science can only be brought about through the positivist method. Hence, as the story goes, Gauss went to the Harz Mountains to measure whether or not space was curved. He selected three peaks as the vertices of a triangle and measured to see whether the three internal angles added up to 180°: if they did, then the space geometry would be Euclidean and not curved; if they did not, then space would be curved. This story is in all probability fictitious. Nevertheless, it is highly philosophic: it tells us that to clarify the finite-infinite issue, we must resort to experiments.

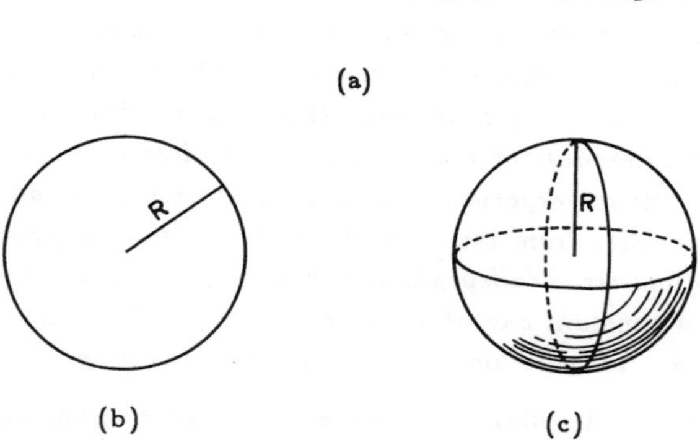

Fig. 4.2. Finiteness does not always imply a boundary; infiniteness does not always imply no boundary. (a) An infinite 1-dimensional space with no boundary: the line; (b) A finite 1-dimensional space with no boundary: the circle; (c) A finite 2-dimensional space with no boundary: the spherical surface.

Fig. 4.3. Gauss went to the extent of surveying a triangle in the Harz mountains formed by Inselberg, Brocken, and Hoher Hagen to see whether the sum of the triangle's interior angles was 180°.

Expansion of the Universe and the Finite-Infinite Issue

Even if Gauss's experiment was performed, it would not have succeeded, because it required high-accuracy measurements which are difficult to achieve even with today's technology. It is only by relying on measurements of a cosmic scale that an experimental approach to the finite-infinite issue is possible.

The universal expansion is the first observed phenomenon on the cosmic scale. What does it say on the finite-infinite question?

At first sight it might seem that if the universe is expanding, then cosmic space should be finite. For expansion of a system means the size of the system is increasing and a system with a size must be finite. Hence for many people, as soon as they hear expansion of the universe, they think the universe is finite, as if the two are synonymous. But this is a misunderstanding.

Finite systems can expand, infinite systems can also expand. Fig. 4.4 shows a one-dimensional, infinite universe, in which each galaxy is represented by an integer. At time t_1, the galaxies are distributed at equal intervals showing the cosmic matter to be homogeneous. If this universe is expanding so that, at time t_2, the original galaxy 1 is now at the position of the original galaxy 2, galaxy 2 now at galaxy 4, and so on, then this is uniform expansion. Before the expansion, there is an infinite space containing an infinity of galaxies, after the expansion it is also the same. This may be called expansion from infiniteness to infiniteness.

The mathematician Cantor developed a mathematical theory dealing with infinities, which can be used to compare various types of infinities. According to his theory, the following two sequences represent the same infinity:

$$n = 1, 2, 3, 4, 5, \ldots$$
$$m = 2, 4, 6, 8, 10, \ldots$$

His reason is this: there is a one-to-one relation between the elements of these two infinite sequences, that is, $m = 2n$, or the elements of m and n can be placed one against one, so neither sequence is more numerous than the other and the two sequences are the same. This argument of Cantor's can be applied word for word to elicit the property of universal expansion: the universe is always expanding while the infinite space always keeps the same infinite character.

In short, merely from the rough idea of universal expansion, we cannot derive anything new concerning the finite-infinite question.

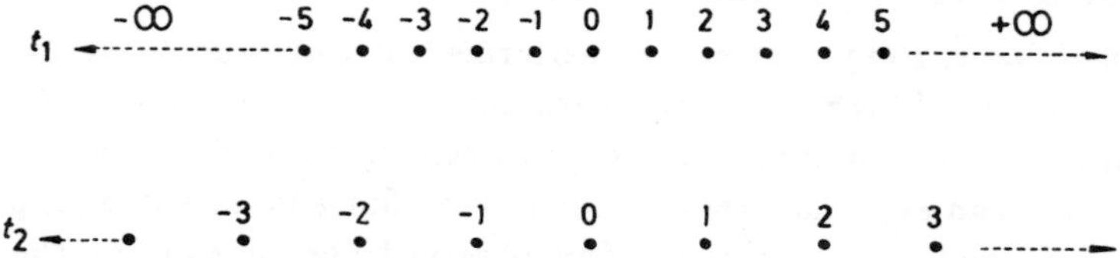

t_1 -∞ -5 -4 -3 -2 -1 0 1 2 3 4 5 +∞

t_2 -3 -2 -1 0 1 2 3

Fig. 4.4. Expansion from infiniteness to infiniteness. Each galaxy in the one-dimensional, infinite universe is represented by an integer.

Prospect of the Expanding Universe

We now proceed to make a quantitative study of cosmic expansion. The basic question is, will the expansion go on forever or will it stop one day and reverse into a universal contraction?

To answer this question, we shall return to the discussion of Chapter 2 and consider the system of galaxies in our surroundings. According to the Copernican Principle, the system constitutes a typical local region. Taking ourselves as the centre, we single out a spherical region with radius R. According to the Hubble relation (2.11), the velocity of galaxy S at the surface of the sphere, relative to the centre is

$$v_S = H_0 R \tag{4.1}$$

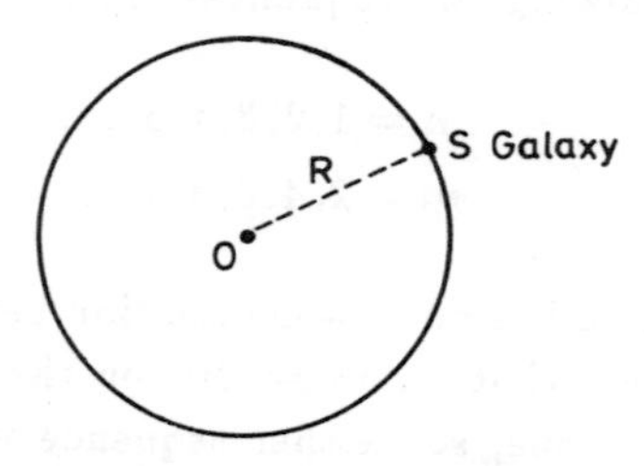

Fig. 4.5. A typically spherical region in the universe with a centre at O and radius R.

Galaxy S is subject to two forces, one is the gravitation of the matter within the sphere, and the other is the gravitation of the matter outside. Because matter is uniformly distributed in the universe, the total gravitation due to the matter outside is zero — this is a basic property of gravitational force, namely, the gravitation of a hollow spherical shell on any object inside the shell is zero. Gravitation due to the matter within the sphere will decelerate the expanding motion of S, just like an object thrown upwards from the Earth's surface is decelerated by the Earth's gravity.

The outcome of the deceleration may be of two kinds. In one, despite the deceleration, S will continue to expand forever; in the other, the deceleration will stop the expansion and change it into a contraction. Which one is in store for S depends on whether or not its velocity v_S is greater than the escape velocity. This is the same criterion that distinguishes artificial satellites and artificial planets. If the spacecraft has an initial velocity greater than the escape velocity then it will escape the Earth and never return; otherwise, it will fall back to the ground after being decelerated.

For the galaxy S, the escape velocity is

$$v_e = (2GM/R)^{\frac{1}{2}} , \tag{4.2}$$

where M is the total mass within the sphere R. It is equal to

$$M = (4\pi/3)R^3 \rho_0 , \tag{4.3}$$

where ρ_0 is the mean mass density of the present universe. If

$$v_S > v_e , \tag{4.4}$$

then galaxy S will expand forever; on the contrary, if

$$v_S < v_e , \tag{4.5}$$

then S will change from expansion to contraction. Using (4.1), (4.2) and (4.3), the conditions (4.4) and (4.5) can be written respectively as

$$\rho_0 < \rho_c \tag{4.6a}$$

and

$$\rho_0 > \rho_c , \tag{4.6b}$$

where ρ_c is called the critical density, defined as

$$\rho_c = H_0^2/[(8\pi/3)G] . \tag{4.7}$$

Thus, the future prospect of S is determined entirely by whether or not the present mass density of the universe ρ_0 is greater than ρ_c.

The future prospect of S is also the future prospect of the entire universe, because S is a typical galaxy and, according to the Copernican Principle, the state of motion of any other region is the same as that of a typical galaxy.

Thus, the future prospect of our universe depends on one parameter, the mean density of cosmic matter.

Mean Mass Density and the Finite-Infinite Issue

The above discussion is based on Newton's theory of gravitation. Strictly speaking, in dealing with cosmological questions, we must use Einstein's theory of relativity.* However, all results are the same whether we use one or the other, that is, all the foregoing formulae will have the same form in relativity.

The two theories do differ greatly on one point, namely, the nature of space. In Newton's theory, the nature of space is absolute: it is not affected in any way by the motion of matter, and is always the infinite space obeying Euclidean geometry. In general relativity, space is not absolute. Its property depends on the motion of matter, that is, a particular state of material motion corresponds to a space with a particular set of features. The two prospects of the universe correspond to two spaces with entirely different properties. Specifically

1. For an ever-expanding universe, the space curvature is negative, the space of the universe is infinite, boundless and open.
2. For a universe that changes from expansion to contraction, the space curvature is positive, the space of the universe is finite, boundless, and closed.

In a book of this size, it is not possible to give a rigorous proof of these two statements, for it requires too much preliminary knowledge in geometry and general relativity. But we can roughly show their rationale. Case 1 corresponds to the case of the galaxy S expanding forever. Such a motion itself will involve an infinite range, therefore, the perpetual existence of expansion indicates an infinite space. Case 2 corresponds to S contracting after expanding, such a motion involves only a finite range in space. Hence such a first-expanding then-contracting universe could be finite.

At this point, we realise that whether the universe is finite or infinite depends only on this one parameter, the mean density of cosmic matter ρ_0.

---

*cf. L.Z. Fang & Y. Q. Chu *From Newton's Laws to Einstein's Relativity*, World Scientific & Science Press, 1987.

If $\rho_0 < \rho_c$, then the universe is infinite,

If $\rho_0 > \rho_c$, then the universe is finite.

An *idiot's* problem, a problem for pure thought, has at last been conquered by science and has become a verifiable problem.

Chapter Five

MATTER VISIBLE AND MATTER INVISIBLE

Density of Galaxies

Whether or not the mean density of cosmic matter ρ_0 is greater than the critical density ρ_c determines whether the universe will contract in the future and whether or not it is finite. Therefore, the measurement of ρ_0 and ρ_c has become a matter of urgency.

According to the definition (4.7) given in the last chapter, ρ_c is

$$\rho_c = H_0^2/(8\pi/3)G \ .$$

Hence it is determined solely by the Hubble constant H_0. From the value of H_0 given in Chapter 2, we have

$$\rho_c \approx 5 \times 10^{-30}\,\mathrm{g/cm^3} \ . \tag{5.1}$$

The measurement of ρ_0 is more difficult, for the universe consists of all sorts of matter and it all contributes to ρ_0. However, we can always start by tackling the easiest part. Seen through the telescope, the sky is a panorama of galaxies. So, we shall first investigate how much galaxies contribute to ρ_0.

If the mean mass of a galaxy is M_G and there are n galaxies in a unit volume on the average, then the contribution is

$$\rho_G = n\, M_G \ . \tag{5.2}$$

Both n and M_G are measurable quantities. For some fixed region of the sky, we count the number of galaxies and find n. As for the galactic mass M_G, this can be measured by using the same method as that for the Sun. We know that the mass of the Sun is measured from the properties of the motion of the planets. For example, the known distance between the Earth and the Sun is $d = 1.5 \times 10^8$ km, and we also know that the time taken for the Earth to go around the Sun is $T = 1$ yr $= 3 \times 10^7 s$. Using these two numbers, we can then derive the mass of the Sun $M_\odot$ as follows

$$M_\odot = \frac{4\pi^2 d^3}{GT^2} \approx 2 \times 10^{33}\,\text{g}\,. \tag{5.3}$$

This formula comes from Newton's theory of gravitation.*

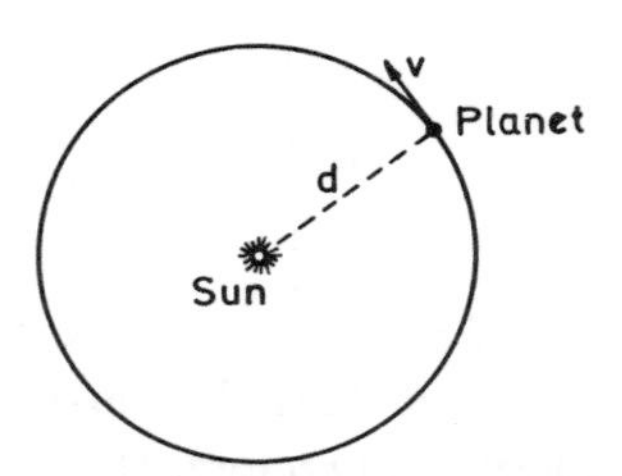

Fig. 5.1. The mass of the Sun can be measured by the periods of a planet's movements around the Sun.

Like the planets in the solar systems, the stars in many galaxies revolve around the centre of the galaxy. Hence, once we know the rotating velocity of some of the stars and their distance from the centre, we can use the above formula to find M_G.

Such measurements show that

$$\rho_G \leq 0.05\rho_c\,. \tag{5.4}$$

If cosmic matter is mainly in galaxies, that is, if $\rho_0 \approx \rho_G$, then we have

$$\rho_0 < \rho_c\,, \tag{5.5}$$

and the universe will be infinite, open and ever-expanding.

*cf. L. Z. Fang & S. X. Li, "Introduction to Mechanics", Science Press, Anhui, 1986.

The Missing Mass

Obviously, matter in the universe is not entirely concentrated in galaxies. The space between the galaxies is not a vacuum but contains, for example, gas or extinguished stars. The question is, just how much such dark matter is there? During the 1930's the Swiss astronomer Zwicky did some work that elucidated the problem of dark matter.

Zwicky's work consists of measuring the mass of clusters of galaxies in two ways. One way is via the luminosity, that is by measuring the luminosities of the member galaxies of a cluster. Because there is a definite relation between the luminosity of a galaxy and its mass, we can deduce the mass of the galaxies from luminosity measurements. Then by adding up all the masses of the member galaxies, we obtain the total mass of the cluster. The other method is based on dynamics, that is, by measuring the relative velocities among the galaxies. Since the mean relative velocity is determined by the mass of the whole cluster, we can also obtain the mass of the cluster from the velocity measurements.

Zwicky discovered that the masses found by these two methods differed greatly. For example, the dynamical mass of the Coma cluster is 400 times the luminosity mass! This result can only be interpreted by the fact that the main mass of the Coma cluster is not contributed by the visible galaxies, but by a large amount of invisible matter within the cluster. The mass measured from the luminosities includes only the mass in light-emitting regions, and does not include the mass of any matter that exists in regions not emitting light. Hence, should the non-light emitting regions contain a large amount of matter, then the luminosity mass would be less than the dynamical mass. As to just what sort of matter contributes to this mass, nothing was known at that time. Therefore, this mass was called "missing mass" or "deficit mass".

Zwicky's bold conjecture has never received public recognition. Even in the 70's, influential opinions still held that galaxies were the main components of the universe, that there was no such thing as the "missing mass", that mass did not have a "deficit" and that the difference between the luminosity and dynamical masses was due to some other cause. The Astronomy volume of the Chinese Encyclopaedia, published in 1980, states the following about the "missing mass": "this problem (referring to the discrepancy between the two masses) has not yet received a satisfactory solution".

Since the 50's, facts supporting Zwicky began to increase, one of which relates to research on the relation between apparent magnitude and redshift.

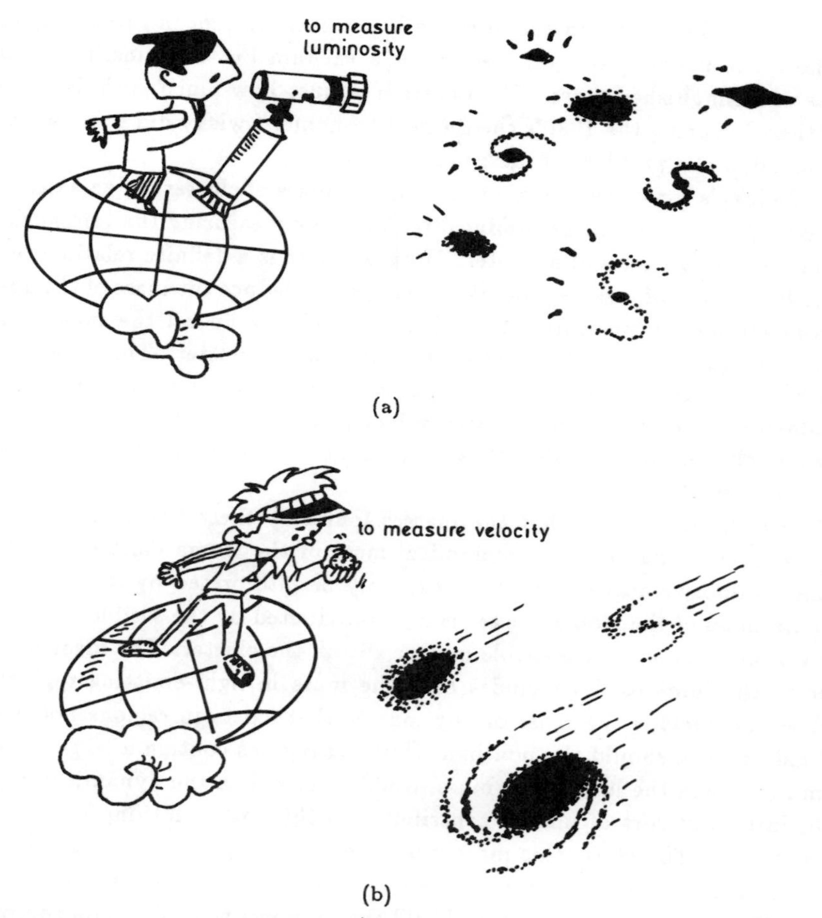

Fig. 5.2. Methods of measuring the masses of galaxy clusters. (a) The *luminosity* method, to measure the mean mass-to-light ratio of galaxies (M/L) and the total luminosity of clusters L_t; the mass is given by $M_c = L_t(M/L)$. (b) The *dynamical* method, to measure the mean velocity of galaxy clusters; the mass of clusters is determined the same way as shown in Fig. 5.1.

The Magnitude-Redshift Relation of Quasars

We refer to Fig. 4.5 of the last chapter, where the galaxy S has a decelerating motion. The size of the deceleration directly determines ρ_0. Therefore, once we measure the deceleration of S, we can calculate ρ_0. Suppose we observe the rate of expansion of the same galaxy S at two different times and see how much it has decelerated to obtain the rate of deceleration. Unfortunately, over human life spans, the change in velocity would be too small and would not be measurable at all. In fact, the observing error would be much greater than the change to be measured. This idea is therefore pointless.

However, since the speed of light is finite, the more distant objects we observe would correspond to earlier times, and the nearer ones, to later times. These two kinds of objects have therefore been decelerated over different time spans. For the nearer objects, the span is longer, hence the deceleration is larger. For the more distant objects, the deceleration is smaller. This is to say that for these two kinds of objects, the ratio between velocity and distance is not exactly the same, i.e., the formula (2.11) is not strictly satisfactory. For the more distant object, the velocity-to-distance ratio is larger; for the nearer ones, it is smaller.

Thus, the relation between distance and velocity is neither in strict proportionality nor a precisely straight line, rather, it is a curve that bends slightly upwards (see Fig. 5.3). The degree to which it bends is determined by the size of the deceleration — the stronger the deceleration, the steeper the curve. Thus, the velocity-distance relation of astronomical objects offers a possibility of determining ρ_0.

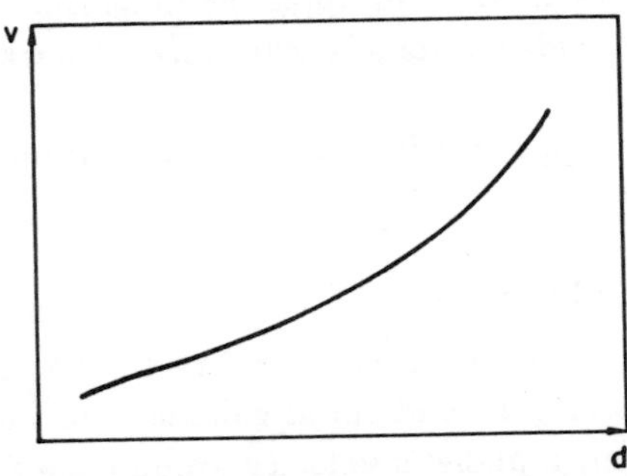

Fig. 5.3. The expansion of the universe is by deceleration, so the relationship of distance and velocity is not a straight line, but a curve that bends slightly upwards.

Generally speaking, more distant objects have greater apparent magnitudes, while velocity is directly deduced from redshift. Therefore, the distance-velocity relation is just the apparent magnitude — redshift relation. The use of this method to find ρ_0 should satisfy two conditions:

1) we should have a class of objects with the same luminosity, as such

objects have a proportionality between the distance and the apparent magnitude.

2) The objects should have large redshifts or large velocities for it is only on diagrams containing large velocities that the upward bend can be seen.

Of the known objects, only quasars have sufficiently large redshifts. Hence, it is very natural to attempt to determine ρ_0 using the apparent magnitude-velocity relation of quasars.

However, among quasars, the luminosity differences are too great for them to be considered as objects of the same luminosity. The present strategy is to select from some fixed standard the quasars that have comparable luminosities and study the apparent magnitude-redshift diagram of this subset. A number of ways in selecting quasars are now available and various quasar subsets have been studied. Almost all such studies have the same result namely that

$$\rho_0 > \rho_c \ , \tag{5.6}$$

or that the universe should be finite, closed and eventually collapsing.

A comparison of (5.4) and (5.6) likewise suggests the existence of dark matter. Although this result also supports Zwicky's, there is a general reluctance to accept the idea of the existence of a large amount of dark matter because the error in the method of apparent magnitude-redshift diagram is considerable. So the evidence for the existence of dark matter is still not all-compelling.

Such evidence first appeared in 1978. It comes from the rotation curves of galaxies.

Rotation Curves of Galaxies

The so-called galactic rotation curve refers to the relation between the velocity of bodies revolving around spiral galaxies and their radius.

For the solar system, a planet's velocity around the Sun v and its radius r are related thus

$$v \propto r^{-1/2} \tag{5.7}$$

that is, the further away a planet is from the Sun, the smaller is its rotational velocity. This is one of Kepler's laws and is applicable to motions around any massive central body. Therefore, if the mass of the galaxy is concentrated in the light-emitting region, then objects outside the light-emitting region must obey the above stated Kepler's Law, with objects further away from the centre rotating more slowly.

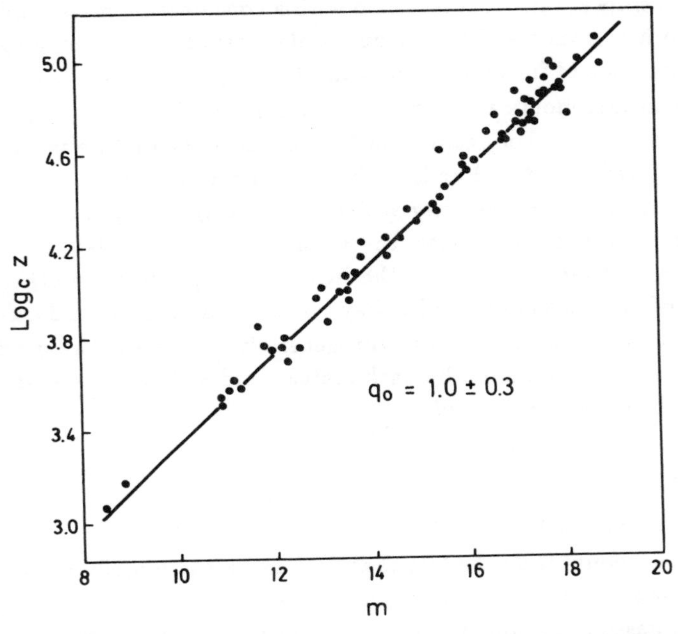

Fig. 5.4. The Hubble diagram for galaxies, where z is redshift and m denotes apparent magnitude. Because the redshifts of galaxies are too small, ρ_0 might not be deduced from this diagram.

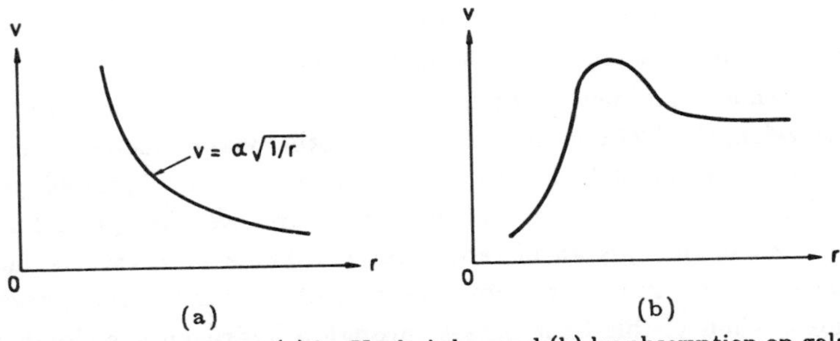

Fig. 5.5. Rotation curves given (a) by Kepler's law and (b) by observation on galaxies, for which the velocity is almost constant.

The observed result is completely different from Kepler's Law (5.7). The observations reveal that outside the light-emitting region of the galaxy, the rotating velocity of objects is independent of their distance. This is to say that objects at different distances have the same rotating velocity! For this "abnormal" result, the only possible interpretation is that the space surrounding the galaxy is not empty; rather, it is a halo with a considerable mass. It does not emit light so it is invisible. This unequivocal evidence caused some to believe that large amounts of invisible or dark matter may exist in the universe.

Subsequent astronomical observations presented more evidence for the existence of dark matter. For example in 1983 a star named R15 was discovered. It was at a distance of two hundred thousand light-years from the centre of the Milky Way, and had a velocity as fast as 465 km/s along the line of sight. To generate such a large velocity, the total mass of the Milky Way system must be least 10 times the mass of the light-emitting region. In other words, nine-tenths of the mass in the Milky Way system is in the form of dark matter.

Today, the following thesis has been generally accepted. The mass of the universe is mainly contributed by dark matter, and *at least nine-tenths* of the matter in the universe is invisible.

Not Baryons

What matter is dark matter?

First, some thoughts on the fact that dark matter is contributed by diffuse gas. We know that the Milky Way system has many gas clouds. Could there also be such gaseous matter in large amounts in the intergalactic space? In fact, we only need to have 1/100 hydrogen atoms in each cubic centimetre inside the cluster of galaxies to give a total that will meet the deficit found by Zwicky. If gas of such a low density is present in a laboratory on Earth, then it can indeed be said to be invisible and hardly measurable. However, in astronomical observations, such a density is quite high.

We all know that neutral hydrogen gas can emit or absorb radio waves with a wavelength of 21 cm and detection or otherwise of such signals will put limits on the density of hydrogen gas. Now, in radio background radiation, no emission lines at 21 cm have been found, nor have absorption lines at 21 cm been seen in the spectra of radio sources. These observations show that the density of hydrogen gas cannot exceed 1/100 atoms per cubic centimeter.

Besides, when visible light passes through a hydrogen gas cloud, absorption in the optical range may also take place. Because the hydrogen atom is a powerful absorber of certain kinds of visible light , the optical method is more sensitive than the radio method. The result is that the density of the intergalactic hydrogen cannot exceed 10^{-12} per cm^3. The optical method can also detect whether any atoms of lithium, carbon, oxygen, magnesium, aluminium, silicon, sulphur or iron are present in the intergalactic space. The result has been entirely negative.

Neither of these two methods can exclude the existence of ionized gas, for ionized hydrogen etc. does not produce 21 cm emission or absorption lines. However, a high-temperature ionized gas will emit X-rays, and X-rays have indeed been found in clusters of galaxies. But the density of ionized gas so derived is again very small, far from enough to account for the deficit mass.

If the deficit mass exists in the form of dust, then that will cause dimming of the starlight. Quantitative estimates gave the result that the mass of diffuse dust can amount to at most 1/100 that of the stars in the clusters, hence, dust cannot be the main component of dark matter either.

We might also guess that the missing mass consists of darkened, "dead" stars or galaxies. Now, if there are so many "dead" stars in today's universe, then at an earlier time, the number of "live" stars must have been much greater than at present. It may seem that astronomical observation has reached "somewhat distant" places, i.e., "somewhat earlier" times because light takes a finite time to reach us. Although we cannot directly see the "distant" stars, such distant objects contribute to the background lights of the sky, and if there were indeed many "live" stars at these distant places, the background radiation of the sky would be much greater than is actually observed. Hence, the idea that there are many "dead stars" in the universe cannot hold either.

From gas, through dust, to "dead" stars: all the different forms of matter that can possibly be formed by chemical elements have one by one been eliminated. We have thus excluded all possibilities of dark matter being some form of baryon matter.

If dark matter is not baryon matter, what else can it be?

List of Candidates

In 1980, some particle physicists declared that the rest mass of the neutrino may be non-zero, and a Soviet experimental group stated more specifically that the rest mass of the electron neutrino is about 6×10^{-32} grams. At that time, this news elicited greater response among astrophysicists than particle physicists. This is because many people thought that neutrinos may just be the missing matter.

Neutrinos are not baryons. The universe contains a vast number of neutrinos. If a neutrino has a tiny rest mass it can far exceed the mass of the baryon component, so becoming the dominant component of the cosmic mass.

More specifically, there are about equal numbers of photons and neutrinos in the present universe, about 400 per cm^3. The average photon mass is 1.1×10^{-36} grams. If the neutrinos have zero rest mass, then their contribution to the mean mass density will be only $1.1\times 10^{-36}\times 400 = 4.4\times 10^{-34}$ g/cm$^3 \ll \rho_c$. If the rest mass of the neutrino, m_ν, is not zero, then the result will be very different. The mass of the neutrino is about 6×10^{-32} grams, and this does not decrease as the universe expands. Hence, the contribution by neutrinos to ρ_0 is

$$6\times 10^{-32}\times 400 = 2.4\times 10^{-29}\ \mathrm{g/cm^3} > \rho_c$$

that is, the rest mass of the neutrino could make the universe finite and closed.

The question of whether neutrinos have a rest mass has still not been settled. Nevertheless, through discussion of the neutrino problem, the entire perspective on research of dark matter has changed. Previously it was regarded as a difficult problem. Now it is regarded as an area to be opened up by the combined efforts of astrophysics and particle physics.

Table 5.1. A List of Dark Matter Candidates.

Name of Particle	Spin	Possible rest mass (eV/c^2)	Present number density limit (cm^{-3})
gravitino	3/2	10^3	1
neutrino	1/2	10	10^2
photino	1/2	10^3	1
gluino	1/2	–	–
Wino	1/2	–	–
Zino	1/2	–	–
Superneutrino	0	10^{11}	10^{-8}
axion		10^{-5}	10^9
GUT magnetic monopole		10^{25}	10^{-22}
KK particle		10^{28}	10^{-25}

Particle physicists hope to find or confirm the many "dark" particles they have predicted. The theories of supersymmetry and supergravity that have emerged have predicted the existence of many new particles. They are not baryons and most of them do not participate in electromagnetic interactions. They have very weak interactions and cannot be detected in today's laboratories. Such features match the properties expected of dark matter. Hence some people have conjectured that the dark matter in the universe consists of just these particles. Table 5.1 is a list of various particles or "candidates" that may possibly be found in dark matter. The table shows that the majority of

Fig. 5.6. Matter in the universe maybe at least 90% invisible.

candidates are "inos". Hence, we may say that more than 90% of the cosmic matter may be inos and not chemical elements.

The question of whether the universe is, after all, finite or infinite seems to be completely dependent on the properties and quantity of inos.

See the Small in the Large, See the Dark in the Bright

The exact properties and quantities of inos are difficult to determine. Astronomical observations only provide data on large and bright bodies in the sky. Can we then deduce from these facts the properties of tiny and dark particles? In brief, we need to find ways by which we can learn about the small from the large and learn about the dark from the luminous.

In this respect, there has been some progress. In the surroundings of our

Milky Way system, there are some 6 or 7 so-called dwarf spheroidal objects; "dwarf", because their luminosities are low, "spheroidal", because their shape is quite spherically symmetrical. These dwarf spheroidal objects are located in the range of 200 to 600 thousand light-years from the Milky Way. Since they are all within the gravitational field of the Milky Way, their masses cannot be too small or else they would have disintegrated under the tidal force of the Milky Way. On the other hand, their low luminosity indicates that they do not contain much luminous matter. This proves that there must be more dark matter in these dwarf spheroidal objects. The particles making up the dark matter must have a fairly large rest mass, otherwise they would not have remained in such "small, dwarf-like" systems. The dark matter probably consists of gravitinos or photinos.

An increasingly popular view is this: The existence of superclusters of galaxies, clusters of galaxies, galaxies and other light-emitting systems in today's universe is directly related to the existence of dark matter. The light-emitting bodies must have a lot of baryons, but the total mass of the baryons is too small and their mutual gravitation is insufficient for them to gather quickly and form those objects. If a large amount of dark matter of the ino type is present, then this difficulty is removed. Since the total mass of the ino component may be ten times that of the baryon component, its gravitation will be a hundred times greater, and this will cause matter to gather quickly to form the objects.

The existence of mankind is based on the premise that stars exist. Without the formation of stars, man could not have multiplied. In this sense, we may say that, although a human being is not made of inos, the present existence of the human species was determined by invisible matter.

If we estimate how much dark matter is most favorable for the formation of the stellar environment of our present lives, then we find that the density of dark matter i.e. the total mass density of ρ_0 should be roughly equal to the critical density ρ_c. The study of dark matter so far has still not enabled us to make a decision on the finite-infinite question. The above argument may be summarized as follows: man is destined by his very existence to face this difficult, finite-infinite question.

Chapter Six

HOW ORDER WAS BORN OF CHAOS

"The World is Getting More and More Complicated"

"The world is indeed getting more and more complicated", many people, when they are feeling low, say with a sigh. This chapter discusses the basis of this true statement.

The things and events we experience through life generally change from the simple to the more complicated. One can easily cite numerous instances in a man's material, social and spiritual life. This observation is not only manifest in human life. Changes in the natural world, too, have this feature of moving from the simple to the complicated.

In Chapter 3, we have explained that all and sundry things in the universe were produced not more than a finite time ago. This "finite time" is, at most, the age of the universe which is 20 billion years. All complicated forms of matter have, without exception, evolved from a simple state in the early universe where no such complications existed.

This is the *evolutionary view* of the universe.

The evolutionary view is commonly held by various civilizations. In "Tao Te Ching", Laotzu wrote: "Tao generates one, one generates two, two generates three, three generates all things". This is the purest expression of evolution from the simple to the complicated. If we stand firm on this view till the end, then we have to maintain that there is only one origin of all things, that all the innumerable things have evolved from this primitive "One". This is also the view of many ancient Greek philosophers. Although they each had their own idea as to what this origin of all things is, they all insisted that there was only *one* origin.

The evolutionary view successfully resolves certain old "paradoxes", like the question of which comes first, the chicken or the egg. According to the evolutionary view, this question is pointless, for neither the chicken nor the egg was there in the beginning. They both evolved gradually from simpler things.

However, the evolutionary view was halted for a long time by a modern paradox — the paradox of thermal death.

The Paradox of Thermal Death

It is generally held that Newtonian mechanics and the theory of Newtonian gravitation are not the theoretical bases of the evolutionary view. Their solutions often serve to explain only cyclic and recurring motions. Cycles and recurrences cannot, of course, be regarded as evolutionary.

The first theoretical basis of the evolutionary view is thermodynamics. According to the Second Law of Thermodynamics, it is almost impossible for Nature to execute cyclic and repeating motions only. Some changes in Nature are irrecoverable and irreversible. Obviously, only irrecoverability and irreversibility can mean development and evolution.

However, the irreversible development and evolution predicted by the Second Law of Thermodynamics does not go from the simple to the complicated as we had hoped. On the contrary, the general tendency of the development of all things in the universe is from the complicated to the simple, from order to chaos, from thermal non-equilibrium to thermal equilibrium.

In Fig. 6.1, a box is divided into two parts A and B. At the beginning, A and B are filled with gases of different temperatures T_1 and T_2, then after A and B interchange heat for a certain time, a common temperature T will eventually be reached. This process is one that cannot be recovered or reversed. That is, the following process does not exist in Nature: Initially A and B contain gases of the same temperature. After a while, a temperature difference is set up between them. In short, a system can evolve spontaneously from a state of thermal non-equilibrium (A and B at different temperatures) to one of thermal equilibrium (A and B at the same temperature), but cannot change in the opposite direction.

Applying the conclusion of thermal equilibrium to the universe, we have to say that the general tendency of the cosmic evolution is for the temperature to be the same all over. In 1854, Helmholtz pointed out in a lecture, that the Second Law of Thermodynamics means that the whole universe will eventually

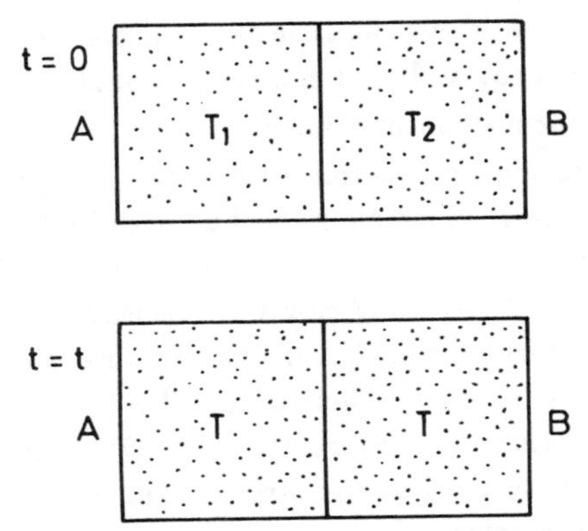

Fig. 6.1. An irreversible process: temperature always evolves from different (t=0) to same (t=t).

be in a state of uniform temperature and "from this point on, the universe will be falling into a state of eternal rest". Thus was born the thesis of thermal death.

The notion of thermal death caused a great stir in Europe a hundred years ago. Not only scientists, but also scholars in humanities discussed it constantly, so much so that one author declared that a scholar totally ignorant of the Second Law of Thermodynamics is as bad as a scientist not knowing a word of Shakespeare.

The difference here is that Shakespeare's work is a symbol of moral aspirations, whereas the Second Law was regarded as the very abyss of decline. For the Second Law predicts that everything will inevitably go from the orderly to the disorderly, from order to confusion, and might this not be the real cause of social regress and moral decline? An "End of the World" pessimism swept over the West. The Chinese, accustomed to desist from the worries of the men of Qi, cared little for such a thing as the "End of the World". But one point must be taken seriously, that is, thermal death and cosmic evolution are directly opposed to each other. The evolutionary view is optimistic, it maintains that Nature always develops in the direction of greater order and more complicated structures, whereas the theory of thermal death maintains that human society together with the entire universe cannot help ending in a tepid silence.

During the last century, a succession of hypotheses was proposed. The hypotheses attempted to show that thermal death is only a paradox, and also to rescue physics from the impossible situation of disintegrating. These attempts were entirely admirable because their futuristic success would mean that physics can be rescued, and that mankind would be saved.

Unfortunately, all such attempts failed.

Failure caused dejection and people vented their anger on thermal death itself. The theory of thermal death had thus acquired a bad name, like a devil that could not be chased away. However, emotions are no substitute for science. Unwelcome theses may not be unscientific theses. As regards the criticisms of the theory of thermal death in some philosophy textbooks, though they commanded respect, when judged by their critical content and psychology, they amounted to no more than Ah Q's "damn it" — a simple invective, which when uttered alone invariably guaranteed the hero's moral superiority over all adversaries.

Fig. 6.2. An unwelcome statement may not be an unscientific statement.

The paradox of thermal death is a deduction from thermodynamics. It can only be resolved by thermodynamics. As the Chinese saying goes, it takes the person who tied the belt to untie it.

Thermodynamics of the Expanding Universe

The key to resolve the paradox of thermal death is cosmic expansion.

As stated earlier for A and B of Fig. 6.1, according to the theory of thermodynamics any temperature differences between A and B will tend toward zero.

Using thermodynamics, we shall prove that for an expanding universe, even if the initial temperature is the same, a temperature difference may still be generated.

Let us return to the typical, spherical region R (Fig. 4.5) discussed in Chapter 4. We assume that the matter in R has already reached equilibrium at the beginning. All forms of matter in it have the same temperature.

Roughly speaking, there are two kinds of matter in the universe, one is the particulate kind of baryons such as neutrons and protons, and the other is radiation, such as light. Let ρ_m and ρ_r be the mass densities of particulate matter and radiation respectively. The energy densities are then,

for radiation $\varepsilon_r = c^2\rho_r$ (6.1)

for particles $\varepsilon_m = c^2\rho_m$. (6.2)

The two kinds of matter have different equations of state,

for radiation $P_r = (1/3)\varepsilon_r$ (6.3)

for particles $\varepsilon_m - nmc^2 = (3/2)nkT$ (6.4)

where P_r is the radiation pressure, n is the number density of the particles, and m is the rest mass of the particles. Here we have assumed that each particle has three degrees of freedom.

The expansion of the universe should be adiabatic. There is no heat exchange with the "exterior" for the "exterior" does not exist in the universe. No other system exists outside the universe. The expansion of the region R, typical of the universe, must also be adiabatic. Even though the "exterior" for region R exists, there is no difference between region R and its exterior because the universe is uniform throughout. There can be no net heat transfer, which is equivalent to being adiabatic. According to thermodynamics, the adiabatic expansion of a system should satisfy the following equation:

$$dE = -P\,dV \tag{6.5}$$

E, P, V being the energy, pressure and volume of the system respectively.

The formulae (6.1)–(6.5) constitute the basis of the thermodynamics of an expanding universe. We now proceed to solve these equations.

Radiation under Adiabatic Expansion

The volume of the region R is

$$V = \frac{4\pi}{3}R^3 \ . \tag{6.6}$$

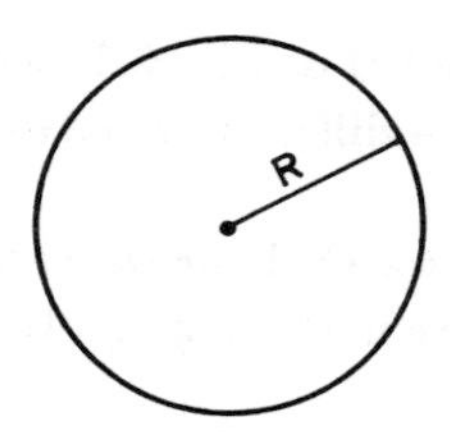

Fig. 6.3. Same as in Fig. 4.5; a typical region in the universe.

The energy of radiation in the region is

$$E_r = \frac{4\pi}{3} R^3 \varepsilon_r \ . \tag{6.7}$$

Substituting (6.6) and (6.7) in (6.5) gives

$$d(R^3 \varepsilon_r) = -P_r dR^3 \ . \tag{6.8}$$

Then, using (6.3), this equation becomes

$$d(R^3 \varepsilon_r) = -\frac{1}{3} \varepsilon_r dR^3 \ , \tag{6.9}$$

or

$$R^3 d\varepsilon_r + \varepsilon_r dR^3 = -\frac{1}{3} \varepsilon_r dR^3 \ ,$$

that is,

$$\frac{d\varepsilon_r}{\varepsilon_r} = -\frac{4}{3} \frac{dR^3}{R^3} \ .$$

The solution of this equation is

$$\varepsilon_r \propto R^{-4} \ . \tag{6.10}$$

According to the thermodynamics of radiation, the relation between the energy density ε_r and the temperature of the radiation T_r is

$$\varepsilon_r \propto T_r^4 \ . \tag{6.11}$$

From (6.6) and (6.11) we immediately get

$$T_r \propto R^{-1} \ . \tag{6.12}$$

This result shows that as the universe expands and as R increases, the radiation temperature falls in inverse ratio to the scale factor R.

Particles under Adiabatic Expansion

We use a similar method to discuss the properties of the particulate matter in the universe. The thermodynamic equation is now

$$d(R^3 \varepsilon_m) = -P_m dR^3 \tag{6.13}$$

where P_m is the pressure of the particulate matter,

$$P_m = nkT_m \ . \tag{6.14}$$

Substituting (6.4) and (6.14) in (6.13), we then find

$$d(R^3 nmc^2) + d(R^3 \frac{3}{2} nkT_m) = -nkT_m dR^3 \ . \tag{6.15}$$

In the expansion process, there will be no increase or decrease in particles. The total number of particles is conserved. Within R, the total number is

$$N = \frac{4\pi}{3} R^3 n \ . \tag{6.16}$$

By conservation of particle number we mean N is a constant, or,

$$dN = d\left(\frac{4\pi}{3} R^3 n\right) = 0 \ , \tag{6.17}$$

or

$$n \propto R^{-3} \tag{6.18}$$

Since m and c^2 are constants, we see from (6.17) that the first term of (6.15) is zero. Thus, (6.15) now becomes

$$\frac{3}{2} d(R^3 n T_m) = -n T_m dR^3 \ ,$$

using (6.17), this becomes

$$\frac{3}{2} \frac{dT_m}{T_m} = -\frac{dR^3}{R^3} \ ,$$

whose solution is

$$T_m \propto R^{-2} \tag{6.19}$$

This result shows that, as the universe expands, the particle temperature T_m also decreases; but the manner of decrease is inversely proportional to the square of the scale factor R.

Generation of Temperature Differences

T_r and T_m vary with R in different ways: one varies in a simple inverse ratio, the other in inverse ratio squared. Hence, during the process of cosmic expansion, it is impossible for T_r and T_m to be always equal. Even if at the beginning we have

$$T_r = T_m \ , \tag{6.20}$$

after expansion we must have

$$T_r > T_m \ , \tag{6.21}$$

for T_r falls more slowly than Tm does. The cosmic expansion causes a state of equal temperatures (6.20) to change into one of unequal temperatures (6.21). This is directly opposite to the prediction of the theory of thermal death. In this manner, the cosmic expansion saves the universe from the final outcome of thermal death.

Those familiar with thermodynamics will certainly ask: the key that led to $T_r \neq T_m$ is to solve the thermodynamic equations separately and independently for the two components, radiation and particles. According to thermodynamics, a system with thermal equilibrium has the same temperature for all the various components, but if radiation and particles always keep the same temperature, how can a temperature difference ever appear?

Those familiar with universal expansion will answer as follows. True, in a system with complete thermal equilibrium all components should have the same temperature. However, a certain time is required to reach equilibrium; interaction between radiation and particles must be carried on for a length of time before the two can achieve the same temperature. If the time required to achieve uniform temperature is longer than the time scale of the cosmic expansion, then there will never be thermal equilibrium between radiation and particles. In this case, it is reasonable to separately solve the thermodynamic equations for the two components. Due to cosmic expansion, the cosmic matter is not in a state of complete thermal equilibrium, rather, it is in one of partial equilibrium. Radiation and particles are separate in thermal equilibrium because there has not been sufficient time for the two components to achieve mutual equilibrium.

The expansion of the universe is based on gravitational interaction. In the last analysis we may say that when we take gravitation into account, we have the possibility of avoiding the conclusion of thermal death. The logical relations of the whole issue are summarized in Table 6.1.

Table 6.1 Theoretical basis for evolution.

Theoretical Basis	Can it explain the evolution from the simple to the complex?
Pure Gravitation	No (cyclic motion)
Thermodynamics without Gravitation	No (thermal death)
Gravitation & Thermodynamics	Possible (avoiding thermal death)

The Negative Thermal Capacity of the Solar System

The intimation Table 6.1 gives us is that the introduction of gravitation into thermodynamics may bring about unexpected results. The paradox of thermal death of the universe has been clarified; now we discuss one of our nearest gravitational systems — the solar system.

For a planet moving in a circular orbit of radius r around the Sun, its velocity is

$$v = (GM_\odot/r)^{\frac{1}{2}} \tag{6.22}$$

where $M_\odot$ is the mass of the Sun. Its kinetic energy is then

$$T = \frac{1}{2}mv^2 = \frac{1}{2}\frac{GM_\odot m}{r} , \tag{6.23}$$

where m is the mass of the planet. It also has a potential energy

$$U = -\frac{GM_\odot m}{r} . \tag{6.24}$$

Therefore, the total energy of the planet is

$$T + U = -\frac{1}{2} \cdot \frac{GM_\odot m}{r} . \tag{6.25}$$

Note that the total energy is negative. Hence, a large r means greater energy, a small r, less energy.

If we give this planet a bit of energy and require it to remain in circular motion, then r must become larger. On the other hand, according to (6.22), if r becomes larger, then the velocity v will become smaller. Hence the net causal relations are,

> *Give energy to a planet and its velocity will become smaller. Take energy from a planet and its velocity will become greater.*

From the thermodynamical point of view "give energy to a planet" corresponds to "transfer heat to the solar system" and "take energy from a planet" corresponds to "the solar system transfers heat to its exterior". Also, "small planetary velocity" corresponds to "low planetary temperature", and "large planetary velocity" to "high planetary temperature", for temperature is directly proportional to the square of velocity. Thus, the above causal relations can be expressed as

> *Add heat to the solar system and the temperature of the system is lowered. Extract heat from the solar system and the temperature of the system is raised.*

In brief, the solar system is such that adding heat lowers the temperature and extracting heat raises the temperature. Its thermal capacity is negative.

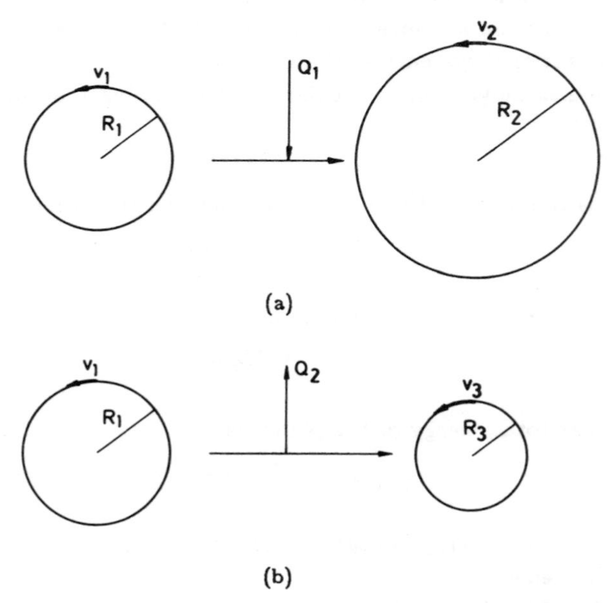

Fig. 6.4. The thermal capacity of a self-gravitational system is negative. (a) Energy is added to the system, i.e. $Q_1>0$, then $v_2<v_1$ or $T_2<T_1$, $\Delta T<0$, so that the thermal capacity $C=Q_1/\Delta T<0$. (b) Energy is removed from the system, i.e. $Q_2<0$, then $v_2>v_1$ or $T_2>T_1$, $\Delta T>0$, so that $C=Q_2/\Delta T<0$.

This startling conclusion applies not only to the solar system but to all systems maintained by gravitation. The thermal capacity of all self-gravitating systems is negative.

No Thermal Equilibrium

The existence of systems with negative thermal capacity is "catastrophic" for thermodynamics.

Consider the system shown in Fig. 6.5, which comprises body a with a positive thermal capacity and body b with a negative thermal capacity. In the beginning, the system is in thermal equilibrium and the temperatures of a and b are equal. The equilibrium is a dynamic one, that is, energy emitted by a is absorbed by b, energy emitted by b is absorbed by a. The two cancel out and equilibrium is maintained.

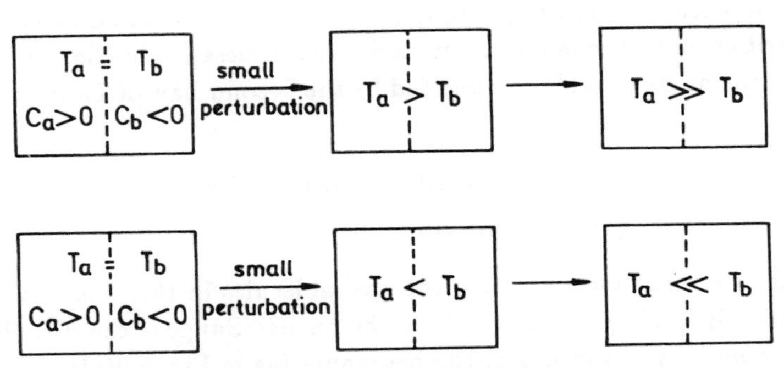

Fig. 6.5. In the case of existence of negative thermal capacity, thermal equilibrium is unstable.

There are always small fluctuations in an equilibrium. For example, the radiation that a gives to b may be slightly larger than what b gives to a, and b absorbs a little more energy. If b has a positive thermal capacity, then b's temperature will rise, its radiation will also increase and so cancel out the excess absorption of energy and return to equilibrium. However, if b has a negative thermal capacity, then an excess in the energy absorbed will lower the temperature, its radiation will become weaker, making it even less able to cancel out the excess given by a. A vicious circle is started, and b's temperature will keep getting lower, destroying the original equilibrium. On the contrary, where the fluctuation is such that b absorbs a little less energy, then the outcome is an ever-increasing temperature of b, again destroying the original equilibrium.

The conclusion is that for the system composed of a and b, thermal equilibrium is unstable and is destroyed by any slight fluctuation giving rise to temperature difference. Hence, as long as self-gravitating systems are present, a stable thermal equilibrium does not exist.

The whole of thermodynamics starts from the existence of thermal equilibrium. For systems in which gravitation plays a decisive role, that sort of thermal equilibrium does not in fact exist. Such systems cannot be in a state of thermodynamic equilibrium, nor in some fixed state differing slightly from equilibrium, rather, they are in unstable states. It is not surprising that certain deductions in thermodynamics do not apply to such states.

Formation of Structures

Let us look at another instructive example for cosmology.

If, in a container of gas, the distribution of the gas molecules is not uniform and has structures (as in Fig. 6.6(a)), then the direction of its evolution is for the distribution to become uniform and structureless (as in Fig. 6.6(b)). This is to say, the mode of evolution decided by the Second Law of Thermodynamics is

$$\begin{aligned} \text{structured} &\longrightarrow \text{structureless} \\ \text{non-uniform} &\longrightarrow \text{uniform}\,. \end{aligned}$$

If the effect of gravitation among the gas molecules in this box of gas cannot be completely neglected, what will be the result? Suppose the distribution of the gas molecules is uniform at the beginning (as in Fig. 6.6(c)). When there is no gravitation, this is the equilibrium state; when there is gravitation, this equilibrium state becomes unstable. As soon as some local region acquires a slightly higher density through fluctuation, its gravitation becomes stronger, attracting more matter, and forming an even greater density. Likewise, if the density in some region is slightly lowered by fluctuation, its gravitation is weakened and more matter will escape, forming a still lower density. In short, a small fluctuation will completely destroy the homogeneous state (see Fig. 6.6(c) & (d)). We therefore see that, in systems with strong gravitation, the direction of evolution is

$$\begin{aligned} \text{structureless} &\longrightarrow \text{structured} \\ \text{uniform} &\longrightarrow \text{non-uniform}\,. \end{aligned}$$

Throughout the universe, gravitation is dominant. Therefore, even if the initial universe is uniform and structureless, it will spontaneously generate a non-uniform and structured state. Clusters of galaxies of various scales owe their formation to this process of inhomogeneity.

At this point, we can answer the question posed at the beginning of this chapter as follows.

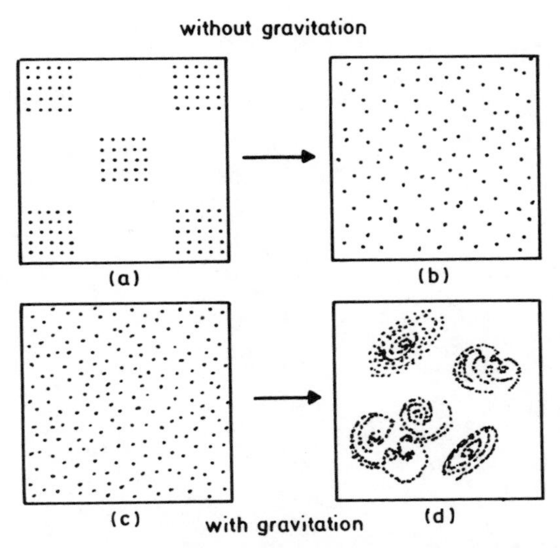

Fig. 6.6. In a system without gravitation, the evolution of the distribution of matter is from non-uniform (a) to uniform (b); in a system with gravitational interaction, the evolution is from uniform (c) to non-uniform (d).

Why is the world getting more complicated? Because there is gravitation.

Why does the simple change into the complex? Because there is gravitation.

Why does chaos become order? Because there is gravitation.

Out of thermal equilibrium, how can thermal nonequilibrium be generated? Again because there is gravitation.

Of course, in addition to gravitation, the universe has to contain different forms of matter like radiation and particles, in order for the above mechanism to operate. In the next chapter, we shall prove that the universe does indeed have the radiation we expect.

Chapter Seven

THE THERMAL HISTORY OF THE UNIVERSE

Message from the Time of Birth of the Universe

The 1978 Nobel Prize in physics was awarded to two American radio astronomers, Penzias and Wilson. In their decision to make this award, the Swedish Academy of Sciences pointed out: "The discovery by Penzias and Wilson is a discovery of fundamental significance: it enables us to obtain information on the cosmic processes occurring at the time of the birth of the universe".

The information discovered by Penzias and Wilson is the *microwave background radiation of the universe.*

This discovery was quite accidental. In 1964, Penzias and Wilson were studying an antenna used in satellite communications at AT&T Bell Laboratories, Holmdel. Holmdel is a small town in New Jersey; although only some two hours drive from New York, it has a completely rural setting and none of the hustle and bustle of the great city. The aim of Penzias and Wilson was to study the radio noise coming from the sky in this quiet place.

To improve the quality of communication, we must seek means to suppress noise. To improve the quality of radio communication, we must reduce the level of radio noise. Radio noise is usually expressed as a temperature, for radio noise is caused by the irregular motion of electrons. The higher the temperature, the more violent the motion, and the greater the noise. There is a definite relation between noise level and temperature. For example, the temperature on the Earth's surface is about 300K, the noise level then is also about 300K.

The antenna used by Penzias and Wilson was a high performance instrument. It had a horn-shaped design. When the horn-opening points to the sky, the 300K ground noise only produces a 0.3K level in the antenna, as compared to a level of 20K–30K in ordinary radio telescopes.

In May 1964, Penzias and Wilson began to use their horn antenna to measure the noise coming from the zenith. The result was 6.7K. After deducting the absorption by the atmosphere and the effect of the antenna itself, 3.5K still came from the zenith. This value was too great, and Penzias and Wilson refused to accept it. They suspected that something was wrong with their antenna. Therefore they checked and rechecked every seam in every metal plate of the antenna. They also cleaned it, but they could not remove this 3.5K. For about a year after this, they found the large noise level not only when the horn was pointed to the zenith, but also when it was pointed into other directions. Furthermore, this noise did not vary with the seasons. It thus became gradually clear that the 3.5K was not produced on the Earth's surface, nor in the solar system, nor in any particular radio source. Rather, it was a diffused radiation throughout the whole space of the universe, *a background radiation.*

Penzias and Wilson did not realize the significance of their discovery. It was not until later, when Dicke and others of the nearby Princeton University told them that this was the very thing the Princeton Group was looking for, that Penzias and Wilson published a short article entitled "Measurement of Excess Aerial Temperature at 4080 MHz" in the Astrophysical Journal. It had little cosmological flavor.

However, the effect of this short article on cosmology can probably be matched only by Hubble's discovery of the redshift. If Hubble's discovery can be said to have opened the door to research on the spacetime structure of the entire universe, Penzias and Wilson's discovery has opened the door to research on the evolution of the nature of matter of the cosmos.

Radiation is Black Body Radiation

Before discussing the theoretical significance of the discovery by Penzias and Wilson, we shall describe their results in more detail.

The first measurements by Penzias and Wilson were made at 4080 MHz or 7.35 cm. Subsequently a series of measurements were carried out in the wavelength range between 0.3 cm and 75 cm. At wavelengths longer than 100 cm, the strong ultra-high frequency radiation from the Milky Way swamps the extragalactic emission and measurements cannot be made. For wavelengths shorter than 3 cm, radiation from the Earth's atmosphere causes trouble and observations can only be made on mountain tops. Besides, only through certain narrow atmospheric "windows" at 0.9 cm, 0.3 cm and so on could radiation from outside the Earth be received. For ranges shorter than 0.3 cm, such "windows" are absent, and measurements can only be made on board high-flying balloons and rockets.

Due to this series of measurements, we can now make a sketch of the spectrum of cosmic background radiation. The result shows that the spectrum is roughly a black body spectrum. (See Fig. 7.1).

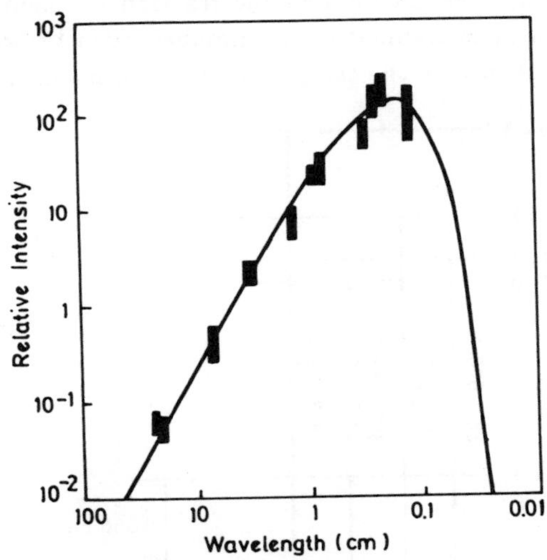

Fig. 7.1. Cosmic microwave background radiation, which possesses a spectrum of 3 K black body radiation.

We know that "black body" radiation is only radiation under the state of thermal equilibrium. If the background radiation has a black body spectrum, it means that the radiation component of the universe is in a state of thermal equilibrium, and we can use a temperature to mark this state. Therefore, an important topic in observational cosmology today is the determination of this temperature.

The value Penzias and Wilson first obtained was 3.5K, with a considerable error. Later measurements were all close to 2.7K. Since ground-based measurements only cover the range between 0.3 cm and 75 cm, they do not reach the peak range for the black body spectrum on the short wave side. (See Fig. 7.1). Therefore, in order to determine more accurately the temperature of the cosmic emission, we must measure the emission intensity at shorter wavelengths.

Using the molecules in the interstellar space, we can indirectly measure the intensity of the background radiation in the shorter wave range. Some molecules in the interstellar space have rather low excitation energies; they are excited by radiation in the millimeter range. For example, as shown in Fig. 7.2 for the molecule cynogen CN, the energy difference between $J = 0$ (base state) and $J = 1$ (first excited state) and that between $J = 1$ and $J = 2$ both fall in the millimeter range, the former at 2.64 mm, the latter at 1.32 mm. If these molecules have reached thermal equilibrium with the background radiation, then an appreciable fraction of them will be in the excited states, and a measurement of the number ratio between the excited and base states will then give the temperature of the background radiation.

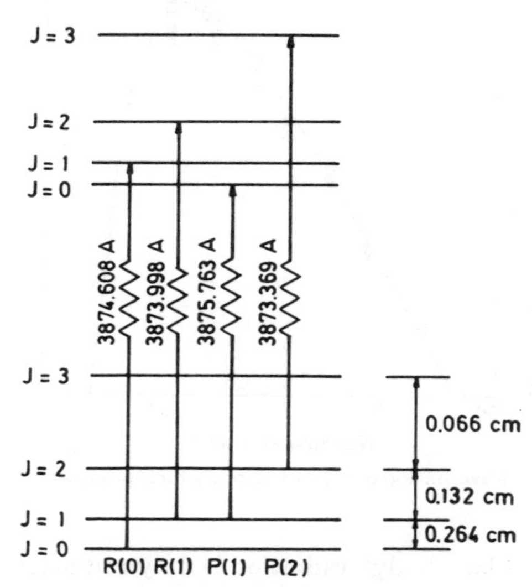

Fig. 7.2. Transitions in the cyanogan absorption spectrum are used to set limits on the temperature of the cosmic microwave background radiation.

The relative numbers of molecules at different energy levels can be measured by optical methods. When starlight passes through the region where these molecules are located, the starlight will be absorbed by the molecules both in the base state and in the excited states. As Fig. 7.2 shows, the absorption spectrum of the CN molecule contains not only the $R(0)$ line, but also the $P(1)$, $R(1)$ and $R(2)$ lines, all of which are in the optically observable range. From the relative intensities of these lines we can deduce the relative distribution of the molecules and hence calculate the temperature of the background radiation.

As a matter of fact, as early as 1941, an investigation of the absorption lines by interstellar CN in the spectrum of the star Zeta Bootes found that CN was excited by a radiation corresponding to 2.3K. At the time the cause was not understood and it was not until the discovery of cosmic background radiation that its significance was realized.

Up to now, the generally accepted background radiation temperature is 2.74K. It is in the microwave range, hence it is also often referred to as the *microwave background radiation.*

The Primeval Fireball

The first inference from the cosmic background radiation is that the early universe must have been very hot.

This point was in fact already discussed in Chapter 1. Because the universal expansion is adiabatic, the temperature of the radiation in the universe is always falling. Hence, the earlier the cosmic time, the higher the temperature. According to formula (6.12), the temperature of the radiation is inversely proportional to the scale factor of the universe:

$$T_r \propto 1/R \tag{7.1}$$

When R is very small, that is, in the early universe, T_r can be very high.

It should be emphasized that T_r is only the temperature of the radiation component. Hence a high T_r is not equivalent to the high temperature of the whole universe. If radiation only accounts for a minute part of the total cosmic content, then even a very high T_r will not matter very much.

According to thermodynamics, the mass density of black body radiation is given by the following formula:

$$\rho_r = a\,T_r^4 \tag{7.2}$$

where the constant

$$a = 8.418 \times 10^{-36}\,\mathrm{g\,cm^{-3}K^{-4}}\ .$$

Inserting $T_r = 2.7\mathrm{K}$ gives

$$\rho_r = 4.47 \times 10^{-34}\,\mathrm{g/cm^3}\ . \tag{7.3}$$

This indeed is a small mass density. Namely, it is only about one ten-thousandth of the critical mass density $\rho_c(\sim 10^{-30})$, and one one-thousandth of the galactic or baryonic density $\rho_G(\sim 10^{-31})$. (cf. Ch. 5). It is indeed a negligibly small component.

Therefore it might seem that we need not be concerned with the temperature of this radiation. However, as shown below, such a statement would be too rash.

Equations (7.1) and (7.2), show that the relation between the mass density of the radiation component and the scale factor of the universe is

$$\rho_r \propto 1/R^4 \ . \tag{7.4}$$

On the other hand, the mass density of baryonic particles is contributed mainly by the rest mass m, that is

$$\rho_m = nm \ , \tag{7.5}$$

where n is the number density of the particles. Because the particle number is conserved in the universal expansion, we have (see (6.18) of Ch. 6),

$$n \propto 1/R^3 \ . \tag{7.6}$$

Hence

$$\rho_m \propto 1/R^3 \ . \tag{7.7}$$

Note this law of variation is different from that for ρ_r (formula (7.4)).

From (7.4) and (7.7), we have

$$\rho_r/\rho_m \propto 1/R \ . \tag{7.8}$$

This can be re-written as

$$\rho_r/\rho_m = (\rho_r/\rho_m)_0(R_0/R) \ , \tag{7.9}$$

where $(\rho_r/\rho_m)_0$ and R_0 represent the present values of the mass density ratio and the scale factor respectively. Formula (7.9) shows that the mass density ratio between radiation and particles is not invariant; rather, it decreases as the universe expands. Although the value of this ratio today is very small, being only

$$(\rho_r/\rho_m)_0 \leq 10^{-3} \ , \tag{7.10}$$

in the early universe, that is when

$$R_0/R \geq 10^3 \ , \tag{7.11}$$

we had

$$\rho_r/\rho_m > 1 \,. \tag{7.12}$$

Namely, radiation was the dominant component of the universe. At that time, its temperature can be regarded as the temperature of the universe.

From (7.1) we have

$$T_r = (T_r)_0(R_0/R) \,, \tag{7.13}$$

where $(T_r)_0 = 2.7\mathrm{K}$ is the present radiation temperature. Accordingly, the condition (7.12) corresponds to

$$T_r > 2.7 \times 10^3 \sim 3000\mathrm{K} \,. \tag{7.14}$$

This is to say, at times when the radiation temperature was higher than 3000K, the universe was filled mainly with radiation, with particles dotted about here and there. This period is called *the radiation of the universe*, or, more popularly, *the state of the primeval fireball.*

Properties of the Radiation Era

During the radiative phase, there were no stars.

Stars are formed when particles gather together through their mutual attraction. The action of radiation is to try to "blow apart" such accumulations. The tails of some comets are formed by the "blowing" action of the scalar radiation "wind", which tries to blow away the cometary particles from the comet. During the radiation era, the radiation wind is much stronger than the gravitation between particles, hence it would blow apart any accumulations.

Therefore, at times when $T_r > 3000\mathrm{K}$, the various material particles of the universe are on the whole uniformly distributed in space, with no accumulation, no clustering, and forming no complicated structures. Also, in an environment hotter than 3000 degrees, many atoms would be ionized. Hence the radiation era was not made of a state of photons plus atoms, rather, it was a sort of uniformly distributed plasma. The plasma was composed of a mixture of electrons, protons, atomic nuclei and photons.

This inference has at least two verifiable predictions:

1) The various types of celestial bodies were all formed at times when $T_r < 3000\mathrm{K}$.

2) The cosmic background radiation distribution should be rather uniform and isotropic.

Prediction 1 agrees well with observations. All the bodies observed so far were formed at those times when $T_r \leq 10\mathrm{K}$. Because of limitations in the present observing power, it is possible that objects are formed during the times of even greater T_r that have not yet been observed. Precisely because of this we are all eagerly awaiting the completion of the sky telescope which will greatly improve the shortcomings of the present observation capability. Once the sky telescope is operational, we shall be able to give a surer answer to the question "At what temperature T_r were the earliest stars formed?"

We now proceed to discuss the result of the observational verification of Prediction 2, or the anisotropy of the background radiation. This calls for a new section.

The Upper Limit of Anisotropy

The problem of anisotropy of the cosmic background radiation can be divided into two aspects. One concerns the anisotropy on small angular scales, the other on large angular scales. The physical implications in these two aspects are not completely the same and there are also differences in the method of their measurements.

If the background radiation is not the relic thermal radiation left over from the time of the primeval fireball but is the result of *superposition* of many discrete radiating sources, then a careful examination will certainly show variations in the radiation intensity on some small angular scales. Scales on which obvious variations are found will be comparable to the sizes of the discrete sources.

Measuring the degree of anisotropy on such small angular scales requires large radio antenna, because the larger the antenna, the greater is its directional resolving power. We fix the antenna on the ground. As the Earth rotates, the antenna points at different parts of the sky. Generally speaking, because the facility is not absolutely stable, the measured noise temperature may slowly drift in time, but such drift will have no great effect on the studies over small angular scales. The slow drift has some small fluctuations. These fluctuations may be due to the anisotropy on a small angular scale, or they may be the noise from the detector itself. Therefore, the measured fluctuation should be regarded as an upper limit for the anisotropy on small angular scales. Table 7.1 lists some early measured results.

Measurements in recent years have further lowered the upper limit. A new result published in 1984 is

$$\Delta T/T < 2 \times 10^{-5} , \qquad (7.15)$$

that is, there is only an anisotropy less than two parts in one hundred thousand. This is very good support for the model of the primeval fireball.

Table 7.1 Small-scale anisotropy of the background radiation.

Wavelength	Angular Scale	Upper Limit of Anisotropy
7.35 cm	40′	2×10^{-3}
3.95	1.4′ × 20′	3×10^{-4}
2.80	10′	2×10^{-3}
0.30	80″	2×10^{-3}

The question of anisotropy on large angular scales has a still deeper significance, which we shall discuss in Chap. 10. Here we shall only talk about the measured results. Measuring large scale anisotropy does not require the use of a large antenna but the effect of the slow drift must be carefully controlled or eliminated. In the earliest measurement, a horn antenna was used to measure the radiation from different directions along the celestial equator and the north celestial pole at intervals of 15 minutes, the difference between the two signals being free from the effect of the drift. Since the north celestial pole has a fixed direction, it follows that if the difference between the two signals varies in time, the background radiation intensity will have different values in different directions. Because of the Earth's rotation, the antenna sweeps through one circuit of the sky every 24 hours. Large-scale variations in radiation intensity should have a period of 24 hours. By isolating the 24 hour period in the measured differential signal, we then derive an upper limit for the large-scale anisotropy. Table 7.2 lists some of the measured results.

Table 7.2. Large-scale anisotropy of the background radiation.

Wavelength (cm)	Angular Scale (degree)	Upper Limit of Anisotropy (10^{-3} K)
4	60	2.3 ± 0.9
3	90	3.3 ± 0.8
0.9 – 1.5	90	3.8 ± 0.3
0.9	90	3.5 ± 0.5
0.3 – 0.5	6	2.9 ± 0.8

It may be seen that over large angular scales, the background radiation also has a high degree of isotropy. This result further proves that the background radiation does not come from any sources within the Milky Way. The reason is that the solar system is located near the edge of the Milky Way system, and any phenomenon in the Milky Way will show a clear anisotropy for the solar system. The high degree of isotropy over small and large angular scales is also a strong support for the Copernican Principle. The distribution of radiation in the universe is homogeneous and isotropic to a rather high degree.

Chronological Table of the Universal Thermal History

The notion of the primeval fireball has received preliminary confirmation. We wish to find further confirmation or rebuttal from more angles.

The most important property of the primeval fireball model is, the earlier the cosmic epoch, the higher the temperature. The temperature is inversely proportional to the universal scale factor. Using this quantitative relation and the solution of universal expansion, we can calculate the temperature at various ages of the universe. Table 7.3 gives the calculated results; it is thus a chronological table of the thermal history of the universe.

Table 7.3. A history of the thermal evolution of the universe.

Age (s)	Temperature (K)	Energy	Main Physical Process
10^{-44}	10^{32}	10^{19} GeV	Quantum gravity
10^{-30}	10^{28}	10^{15} GeV	Particle processes
10^{-12}	10^{16}	10^{3} GeV	
10^{-4}	10^{12}	10^{2} GeV	
1	10^{10}	1 MeV	Nuclear processes
10^{2}	10^{9}	0.1 MeV	
10^{12}	4×10^{3}	0.4 MeV	Atomic processes
10^{12-16}			Gravitational processes
8×10^{17}	2.7	3×10^{-4} eV	

Four items are listed in the table. Column 1 gives the age of the universe; column 2 gives the corresponding temperature T, that is, the radiation temperature T_r; column 3 gives the energy ε corresponding to the temperature, that is,

$$\varepsilon = kT \ ; \tag{7.16}$$

and the last column is the main physical process, which is determined by the energy scale at the time.

We all know that each physical process has its characteristic energy scale. Generally speaking, the stronger the action that sets off the process, the greater the energy scale. The action of gravitational processes is the weakest, and the corresponding energy scales are the smallest; next come the atomic processes, with somewhat larger corresponding energy scales; the nuclear processes are even stronger, with even larger scales; lastly, particle processes are the strongest, and their energy scales are one notch up compared to those of nuclear processes.

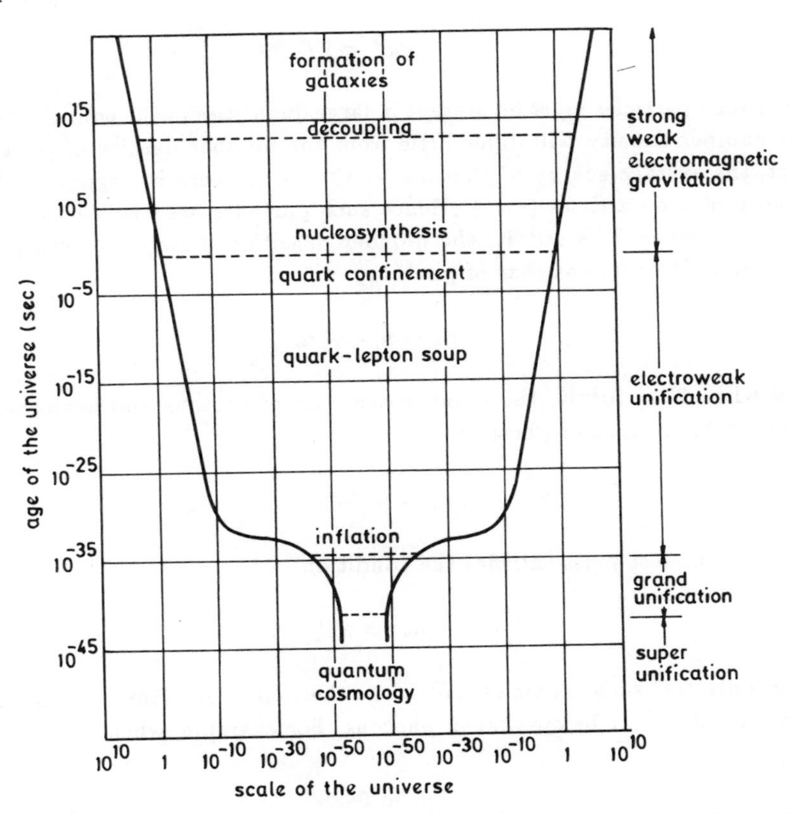

Fig. 7.3. The evolution of the universe. The vertical and the horizontal axes denote the scales of cosmic space and time respectively.

The direction of cosmic evolution is towards lower and lower energy scales. Hence, the earlier the cosmic epoch, the higher the corresponding energy scale. Thus, a chronological ordering of the main physical process in the universe is, by and large, an ordering from stronger to weaker interactions.

The stage of gravitational processes in Table 7.3 has already been discussed in the last and present chapters. During this stage, the main mediating agent among cosmic matter is gravitation and the main process is the clustering of various types of stellar objects from a state of uniform distribution.

In the early universe the cosmic matter was in a state of thermal equilibrium and it is easy to judge which particles were present in such an equilibrium with temperature T. There are three criteria:

1. If the rest mass m of the particle satisfies the condition

$$mc^2 < kT \tag{7.17}$$

then such particles must be present in large quantities in the equilibrium state. Its number density will differ little from the number density of photons. In fact, the average energy of photons ($\sim kT$) in this case is larger than the rest energy of the particles (mc^2). Hence such particles are easily produced. For example, when $T > 10^{10}$K, the number densities of electrons and positrons are about the same as that of photons,

$$n_e \sim n_{e+} \sim n_r \ , \tag{7.18}$$

and when $T > 10^{13}$K, the number densities of protons and antiprotons are comparable to that of photons,

$$n_p \sim n_{\bar{p}} \sim n_r \ . \tag{7.19}$$

2. If the rest mass satisfies the condition

$$mc^2 > kT \ , \tag{7.20}$$

then only few such particles will be in thermal equilibrium. Their number density will be far below that of photons. For example, when $T < 10^{13}$K, we have

$$n_p, n_{\bar{p}} \ll n_r \ ,$$

and when $T < 10^{10}$K, we have

$$n_e, n_{e+} \ll n_r \ ,$$

and the protons and antiprotons of the early stage, as well as the electrons and positrons, will all have been annihilated.

3. Some particles apparently can exist in large quantities even under the condition $mc^2 > kT$. In general, these are particles of the weak or super-weak interactions, and they do not easily vanish through annihilation. In fact, such particles no longer participate in the thermal equilibrium. The invisible particles discussed in Chap. 5 are probably such particles.

The stage of gravitational processes, when $T < 3000\text{K}$, has been discussed in the last chapter and the first section of this chapter. During this stage, the main interaction among cosmic matter is gravitation, and the main physical phenomenon is the change from a homogeneous state into a non-homogeneous, star-studded state. It is the time of star formation. The theoretical predictions for this stage are in overall agreement with the observed results.

We see from Table 7.3 that, for further checks and verification, we should look into cosmic sources of higher energies, that is, the nuclear and particle stages. This will be the topic for the next three chapters.

Chapter Eight

SYNTHESIS OF ELEMENTS

Abundance of Elements — Once Again

We have already talked about the abundance of elements in Chapter 3. Here we make some further remarks.

There are over 90 naturally occurring chemical elements. Their abundance in Nature varies. Some elements are abundant, others are very rare. Geoscience started to measure the percentages by mass of the various elements contained in different samples long ago. These percentages are what we call *the abundance of the elements.*

The aim of the geoscientist in studying the abundance of elements is to clarify the cause of various geological strata. Different geological processes have a different abundance.

Astrophysicists also study the abundance of elements in various heavenly objects. This aim, partly, is the same as in geoscience, to clarify the cause and origin of the various heavenly objects, and partly, to study the origin of the elements themselves.

These two aims are two ways of looking at the subject matter. When we wish to investigate the cause of different heavenly objects, we investigate the difference among the abundances and assume it is due to the different origins of objects. When we wish to inquire into the cause of the elements themselves, we consider the "common properties" or features that are common to different objects or which have the same average results over large scales.

Figure 3.3 of Chapter 3 gives abundances of various elements. From the data, we note the following features:

1. Hydrogen and helium are the most abundant elements. Their abundance adds up to about 99%, which means that all the other elements together amount to only about 1%.

2. In general, the heavier the element, the scarcer it is. Of course, there are some exceptions, like the abundance of iron, which is slightly higher as shown by the small peak in the measurement.

These are the two points that any theory on the origin of elements should be able to explain quantitatively.

Nucleosynthesis in the Stars — Once Again

The theory of nucleosynthesis in the stars as the origin of elements was already described in Chapter 3. Here we make some additional remarks.

In 1938, the nuclear physicist cum astrophysicist Gamow organized a physics symposium in Washington D.C. Among the participants were nuclear physicists as well as astrophysicists. A free exchange of ideas took place between these two fields. H. Bethe, a young nuclear physicist at the time, was present. Inspired by the symposium, Bethe put forward a theory regarding the energy source of the stars – *the theory of thermonuclear fusion.* This marked the inauguration of the theory of elemental evolution.

Bethe was later awarded the Nobel Prize in physics for this theory. The theory says that the source of energy for the Sun or stars comes from nuclear reactions in the core region. Because the central temperature inside the Sun and stars is extremely high, the particles there have very high kinetic energies, and the atomic nuclei can overcome their mutual electrostatic repulsion and collide. The result of collision is the synthesis of heavier nuclei from lighter ones, that is, the thermonuclear fusion. The energy released in the fusion reactions maintains the high temperature of the solar interior and the solar radiation. Specifically, the energy source for the Sun is provided by the so-called proton-proton process. This process includes the following main reactions:

$$p + p \longrightarrow {}^2\mathrm{H} + e^+ + \bar{\nu}_e \ ,$$
$${}^2\mathrm{H} + p \longrightarrow {}^3\mathrm{He} + \gamma \ ,$$
$${}^3\mathrm{He} + {}^3\mathrm{He} \longrightarrow {}^4\mathrm{He} + 2p \ ,$$

where e^+ is a positron, $\bar{\nu}_e$ is an antielectron neutrino, γ is a photon and ${}^2\mathrm{H}$ is deuterium. It may be seen that protons are constantly changed into helium in the Sun.

When the protons are exhausted in the centre of the star, helium atoms fuse to become lithium, lithium fuses to become beryllium, beryllium fuses to become boron, and so on. Therefore, seen from the viewpoint of elemental evolution, stars are constantly changing protons into heavier and heavier nuclei. Studies developed around this topic have formed an independent branch in astrophysics – *nuclear astrophysics*. Its aim is to study how the various elements are produced and transformed in stellar activities.

After the Second World War, many physicists who took part in the making of the atomic bomb turned to astrophysics. Between the nuclear processes of the atomic bomb and those in stars there is not essential difference. As a result, nuclear astrophysics developed rapidly and by 1951 had considerably matured and could offer a plausible interpretation of the entire abundance curve. For each element the stellar nuclear process responsible for its production was found. Today's astrophysicist is able to state what types of stars and what processes have produced the carbon, oxygen, and iron that are in your body.

However, the success of the theory on stellar nucleosynthesis is not complete. As pointed out in Chapter 3, it cannot explain the abundance of some of the lightest elements, particularly helium and deuterium.

We now turn to this question.

The Abundance of Helium

On Earth, helium is scarce; in the stellar realm, its abundance is the second highest. Helium was first discovered in the spectrum of solar prominence and later found on the Earth's surface.

The most remarkable feature about helium is that its abundance, Y, is roughly the same in many different types of objects, being in the range

$$Y \sim 0.25\text{–}0.27 . \tag{8.1}$$

The implication of this result is that the distribution of helium on a large scale is highly uniform.

The most commonly used method for measuring helium abundance is by means of spectra. From the relative intensity of hydrogen and helium lines in the spectrum, we can calculate the relative abundance between the two. All galaxies have some regions of ionized hydrogen, and because the temperature is high, emission lines of both hydrogen and helium can be observed.

Table 8.1 lists the measured results in some galaxies. When the ionized gas is very thin, we can also make a comparison with the radio spectra of hydrogen and helium. The result is similar to what is obtained optically.

Table 8.1 Helium Abundance in Galaxies.

Galaxy	Y
Milky Way	0.29
Small Magellanic Cloud	0.25
Large Magellanic Cloud	0.29
M33	0.34
NGC 6822	0.27
NGC 4449	0.28
NGC 5461	0.28
NGC 5471	0.28
NGC 7679	0.29

The Sun's surface has a temperature of 6000 degrees; it is comparatively cool, so under general conditions, helium lines cannot be seen. They can only be seen in solar prominences. The solar helium abundance estimated from the spectrum of prominences is $Y \sim 0.38$. On the other hand, the Sun is continually emitting outward particle streams, forming the so-called solar wind. These particle streams constitute low-energy cosmic rays, whose abundance give an indication of the solar abundance. When the Sun is quiet, the result obtained this way is $Y \sim 0.20$.

Some may ask, since the result obtained from the spectrum or cosmic rays can only refer to the surface of the star, how can we be sure that the value of Y in the interior is also about 0.25? Obviously the abundance in the interior is more pertinent.

For the stellar interior, we can use an indirect method to find the abundance. The Hertzsprung-Russell diagram of globular clusters was outlined in Chapter 3. From the position of its turn-off point, we can find the age of the cluster. Also, from the overall shape of the diagram we can deduce the hydrogen abundance X, the helium abundance Y and the combined abundance of all the other elements, Z. This is possible because for a given set of values of X, Y, Z and the age, the theory of stellar structure will give a calculated H-R diagram, and that calculated diagram which agrees best with the observed diagram then determines X, Y, Z. The result found this way is $Y \simeq 0.24$–0.33, in agreement with the result from the spectrum method.

The method of stellar structure can also be applied to individual stars. Once we know the mass, luminosity, age, and the abundance of heavy elements Z of the star, we can deduce Y from the theory of stellar structure. Of all the stars we know the Sun best. The Sun's mass is $M_\odot = 1.989 \times 10^{33}$ g, and its luminosity is $L_\odot = 3.83 \times 10^{33}$ erg/s. From the absorption spectra of hydrogen and heavy elements in the solar spectrum, we can estimate that $Z/X \sim 0.019$. As for the Sun's age, we already have an accurate determination of 4.6 billion years. All four required values are given. Accordingly, we can calculate the solar helium abundance and we find that $Y \simeq 0.27$–0.32. This result falls between the prominence and cosmic ray measurements, and is thus highly satisfactory.

The Helium Problem

The greatest shortcoming of the theory of stellar nucleosynthesis is its inability to explain why $Y \sim 0.25$.

We can demonstrate this "inability" in a very simple way. First, each fusion of four protons into one ^{4}He releases an energy of

$$\Delta E = 27 \text{ MeV} . \tag{8.2}$$

In the stars, this amount of energy is used to maintain the starlight as almost all ΔE is released in the form of radiation. Thus, the formation of every ^{4}He is accompanied by the production of a mass of 27 MeV/c^2 in the form of radiation. The mass of one ^{4}He is

$$m_{^4\text{He}} = 3728 \text{ MeV/c}^2 .$$

If all the ^{4}He was synthesized in stars, then the ratio between ρ_r, the mass density of radiation in the universe, and ρ_{He}, the mass density of ^{4}He, should be greater than 27/3728, ("greater than", because the universe must be expected to have other radiation in addition to the above source):

$$\rho_r/\rho_{\text{He}} > 27/3728 \sim 0.007 . \tag{8.3}$$

Besides, from $Y \sim 0.25$, we have

$$\rho_{\text{He}} \sim 0.25\rho_G \tag{8.4}$$

where ρ_G is the mass density of galaxies, or the mass density of baryons. From (8.3) and (8.4) we have

$$\rho_r/\rho_G > 0.002 \tag{8.5}$$

If we take $\rho_G \sim 10^{-31}$ g/cm^3, then we have

$$\rho_r > 2 \times 10^{-34} \text{ g/cm}^3 \tag{8.6}$$

The mass density of radiation can be deduced by observing the background light of the sky. Figure 8.1 shows the intensity of the background emission in various wavelength ranges. Table 8.2 lists the ρ_r values at different wavelengths. The 2.7K background radiation in Fig. 8.1 and in Table 8.2 is familiar to us; as pointed out in the last chapter, this part of the background radiation is not produced by stars, rather, it is the remains of the primeval fireball of the universe. Table 8.2 shows that the mass densities in the other wavelength ranges are all very small, and cannot satisfy the inequality (8.6), even when combined. There is at least a factor of 10 between the observed result and the lower bound in (8.6). We have thus shown that most of the helium in the universe could not have been produced by nucleosynthesis in the stars.

Thus, stellar nucleosynthesis cannot explain the origin of helium.

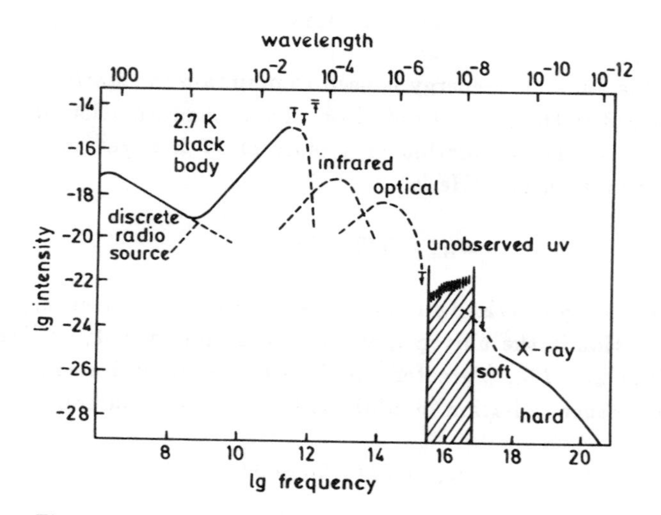

Fig. 8.1. Cosmic background radiation at various wavelengths.

Table 8.2. Mass density of radiation at different wavelengths.

Wavelength Range	Mass Density (g/cm^3)
Radio	$\sim 10^{-40}$
Microwave (2.7K)	4.4×10^{-34}
Visible Light	$\sim 10^{-35}$
X-ray	$\sim 10^{-37}$
γ-ray	$\sim 10^{-38}$

The Primeval Nucleosynthesis

The helium problem finds its solution in cosmology.

According to the chronological table of the thermal history of the universe (Table 7.3) given in the last chapter, the interval between the cosmic ages of 1 second and 100 seconds is one in which the energy scale coincides with that of the nuclear processes. This is another suitable environment for nucleosynthesis. The nucleosynthesis that took place during these two or three minutes is called the "primeval nucleosynthesis".

According to the three criteria given in the last chapter, when the cosmic age $t \sim 1$ second, the cosmic temperature $T \sim 10^{10}$ K, energy $kT \sim 1$ MeV, the universe has large quantities of electrons and positrons, because their rest-mass is $m_e c^2 \sim 0.5$ MeV. At that time, there could not be many neutrons (n) or protons (p), because their rest-mass is $m_p c^2 \simeq 1$ GeV. In such an environment, no atomic nuclei could exist either, because the temperature is so high that all nuclei would be broken into neutrons and protons, just as a high temperature will ionize all atoms into electrons and nuclei.

Even though neutrons and protons are few in number and direct collisions among them do not occur easily, the following processes do occur frequently because of the large number of electrons and positrons:

$$p + \bar{\nu}_e \rightleftharpoons n + e^+ \ , \tag{8.7}$$

$$p + e^- \rightleftharpoons n + \nu_e \ . \tag{8.8}$$

The effect of these processes causes the neutron and proton numbers to reach thermal equilibrium, that is, we shall have

$$n_n / n_p \sim e^{-(m_n - m_p)c^2/kT} \ , \tag{8.9}$$

where n_n and n_p are the neutron and proton number densities respectively, and m_n and m_p are their respective rest-masses. Because $m_n > m_p$, the neutron number is always slightly less than the proton number.

After the universe has expanded further and the temperature has fallen below 10^{10} K, electrons and positrons will no longer exist in large quantities, positrons having been annihilated. Then, the processes (8.7) and (8.8) will no longer occur with ease, and neutrons and protons will cease to be in equilibrium. Their number density ratio will no longer vary with temperature, rather, it will be frozen out at the value for the $\sim 10^{10}$K state.

The mass difference between the neutron and the proton is

$$(m_n - m_p)c^2 = 1.3 \text{ MeV} \ . \tag{8.10}$$

Substituting this value and $T = 10^{10}$K in the formula (8.9), we have

$$n_n : n_p \simeq 1 : 5 . \tag{8.11}$$

This is the number ratio between neutrons and protons after being frozen out.

When the temperature falls further, so $T \simeq 10^9$, the neutrons and the protons begin to fuse into other nuclei. The first fusion process is the formation of deuterium (^{2}H or D):

$$n + p \longrightarrow {}^2\mathrm{H} + \gamma . \tag{8.12}$$

Because the binding energy of deuterium is only 0.26 MeV, it can exist only after the temperature has fallen to 10^9K. When $T > 10^9$K and $kT > 0.26$ MeV, any deuterium will be broken up into a neutron and a proton under the action of photons. This is the reason why nucleosynthesis can only begin when $T \leq 10^9$K.

Once deuterium exists, neutrons and protons will quickly synthesize into helium, ^{4}He, for the following processes are very fast:

$$\begin{aligned} n + p &\longrightarrow {}^2\mathrm{H} + \gamma , \\ {}^2\mathrm{H} + {}^2\mathrm{H} &\longrightarrow {}^3\mathrm{He} + n , \\ {}^2\mathrm{H} + {}^2\mathrm{H} &\longrightarrow {}^3\mathrm{H} + p , \\ {}^3\mathrm{H} + {}^2\mathrm{H} &\longrightarrow {}^4\mathrm{He} + n . \end{aligned}$$

These processes will not cease until all the neutrons have been incorporated into a helium nuclei. The remaining protons will no longer find any neutrons to fuse with, and will thus become the hydrogen found in Nature.

Since each helium (^{4}He) contains 2 neutrons, the number of ^{4}He formed in a unit volume is $n_n/2$. Also each ^{4}He has a mass of 4, hence the helium abundance should be

$$Y = \frac{(n_n/2) \times 4}{n_n + n_p} = \frac{2(n_n/n_p)}{1 + (n_n/n_p)} . \tag{8.13}$$

Substituting the value (8.11) of n_n/n_p above gives

$$Y = 0 : 33 . \tag{8.14}$$

This is the helium abundance produced in the nucleosynthesis of the early universe.

Effect of Neutron Decay

$Y \sim 0.33$ is fairly close to the observed result $Y \sim 0.25$, but we wish to find a better theoretical value.

In the simple calculation above, we neglected the important factor that free neutrons are unstable. They can decay into protons:

$$n \longrightarrow p + e^- + \bar{\nu}_e \ . \tag{8.15}$$

The lifetime of decay is about 10 minutes.

When neutrons and protons cease to be in thermal equilibrium, that is, when $T < 10^{10}$K, protons can no longer change into neutrons. Thus, the process (8.15) becomes irreversible. The neutron decay into protons makes the ratio n_n/n_p smaller and smaller when $T < 10^{10}$K. Between the ceasing of thermal equilibrium at $T \sim 10^{10}$K and the onset of nucleosynthesis at $T \sim 10^9$K, 100 seconds of cosmic time will have elapsed. This time interval of about 2 minutes is not entirely negligible compared to the lifetime of 10 minutes. That is, a small portion of the neutrons will have become protons, making the ratio n_n/n_p fall from 1/5 to

$$n_n/n_p = 1/7 \ . \tag{8.16}$$

Substituting this value into (8.13) immediately gives

$$Y \sim 0.25 \ . \tag{8.17}$$

Its agreement with the observed result given in (8.1) is very satisfactory.

In the above calculation, there was a rather happy combination of two circumstances. One is the lifetime of 10 minutes for the neutron; the other is the time of 2 minutes between $T = 10^{10}$K and $T = 10^9$K. If these values were a little different, then the result would be unrecognizable.

The "10 minutes" is determined by the nuclear decay.

The "2 minutes" is closely connected with the temperature of the background radiation. For example, if the present temperature was not 3K, but slightly higher, then $T \sim 10^{10}$K would not correspond to the cosmic age of 1 second, but to a later time. Since the cosmic expansion is decelerating, the time taken for T to fall from 10^{10}K to 10^9K would be longer than 2 minutes, and this would cause a greater decrease in n_n/n_p. Y would also fall.

It may therefore be seen that the background radiation temperature of 3K and the helium abundance of $Y \sim 0.28$ are mutually dependent; if one changes, the other must also change.

Gamow, whom we mentioned earlier, noticed this interdependence already in the late forties, well before the discovery of the cosmic background radiation. He and others used the value $Y \sim 0.25$ to make the prediction that a cosmic background radiation of about 10K should exist. Unfortunately, at that time, cosmology of the young universe (with an age of a few minutes) was still regarded as closer to fantasy than to science, and Gamow's prediction drew no responses and was soon forgotten. Apparently, after more than ten years, his work had to be done all over again.

Deuterium

The natural abundance of deuterium in Nature is very small, yet it has important cosmological significance.

Deuterium is lively by nature; it participates easily in nuclear reactions. As a result no deuterium could have remained in the stars. Any deuterium there would have "burned out"; just as inside a stove, no petrol would have been left over, only slag. Therefore, it is impossible for the theory of stellar nucleosynthesis to explain why there is any deuterium in Nature since its abundance is ever so small.

The primeval nucleosynthesis again can explain the existence of deuterium. Figure 8.3 shows the variation of the helium and deuterium abundance in primeval nucleosynthesis. It shows that a small amount of deuterium can indeed be left over in this process. Figure 8.4 also illustrates some features of primeval nucleosynthesis, including the abundance of the various elements so produced. The vertical axis is abundance and the horizontal axis is ρ_G, the present mass density of baryons. The deuterium curve depends on ρ_G. Hence, with an accurate measurement of the deuterium abundance we can get an estimate of ρ_G and see whether this estimate agrees with values obtained by other methods. This is also one way of checking the theory. Hence, the measurement of the deuterium abundance is extremely valuable.

In sea water the number ratio between hydrogen and deuterium is 6600 to 1, which is very different from the ratio in the Earth's crust. Neither value can be regarded as the mean abundance of deuterium.

For the solar system, Apollo's finding is of importance. Apollo's lunar-landing brought back an aluminium foil with ions of the solar wind from the moon. From the helium-3 (^{3}He) collected, we can estimate the deuterium abundance. According to our belief, most of the deuterium in the primitive solar nebula has become ^{3}He. The hydrogen/deuterium ratio of the formation of the solar system obtained this way is 40000/1. Also, the result of this observation of Jupiter is 48000/1.

(a)

(b)

Fig. 8.2. (a) Gamov predicted, in the late forties, that a 10 K background should exist in the universe. (b) At that time only few people believed him.

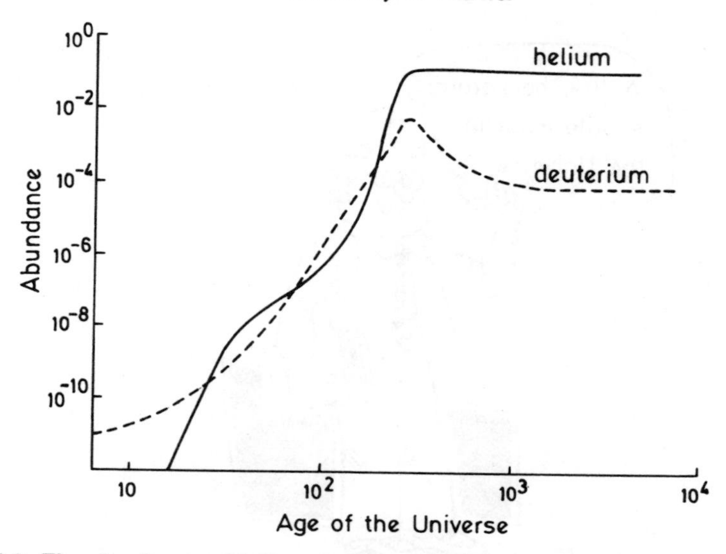

Fig. 8.3. The abundances of helium and deuterium as a function of cosmic time at nucleosynthesis.

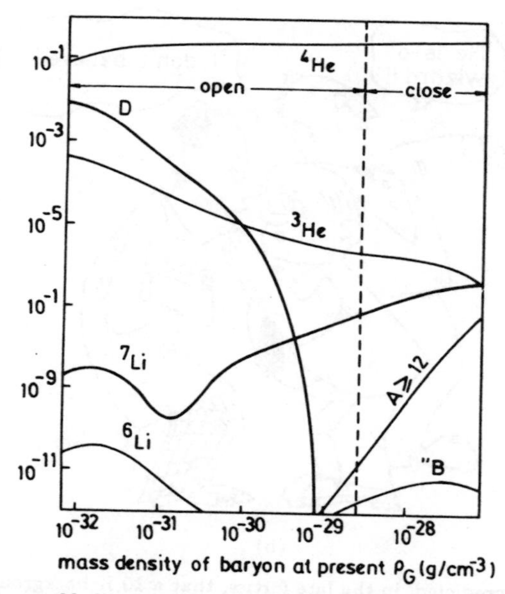

Fig. 8.4. Abundances of helium, deuterium, lithium and other light elements as a function of present baryon density.

Deuterium in intergalactic gas can be measured by methods of radio astronomy, for it has a characteristic spectral line with a wavelength of 92 cm. This line was indeed observed in 1972 in the direction of the galactic centre, and the estimated ratio is between 3000/1 and 50000/1.

The most accurate measurement was made in recent years by the Copernicus satellite. The satellite can observe spectral lines in the ultraviolet part of the spectrum. In particular, using ultraviolet lines we can distinguish between hydrogen cyanide and deuterium cyanide — the two differ only in the substitution of a hydrogen atom by a deuterium atom. The ratio measured this way is between 5000/1 and 500000/1. The Copernicus satellite was also used for a direct measurement of the absorption lines of the deuterium atom. The result was 70000/1.

Averaging the different values above, the present acceptable value for the deuterium abundance is

$$X_b \sim 2 \times 10^{-5} \ .$$

The baryon mass density derived from this is

$$\rho_G \sim 6 \times 10^{-31} \ \mathrm{g/cm^3} \ .$$

This result is in agreement with the results from direct measurements given in Chapter 5. It shows that the theory of primeval nucleosynthesis of the origin of light elements is highly successful.

Chapter Nine

ORIGIN OF ASYMMETRY

Dirac's Error

In 1933, Dirac was awarded the Nobel Prize in physics for the success of his theory on the electron and the positron. In the awarding ceremony in Stockholm on December 12, he gave a lecture containing the following passage which, seen today, is largely incorrect:

> *"If we accept the view of complete symmetry between positrons and electrons as revealed in the fundamental laws of Nature, then we must see the following circumstance as an accident: on Earth, and possibly in the whole solar system, electrons and positive protons predominate numerically. It is very probable that this is not the case with some stars, that these stars are mainly made of positrons and negative protons. In fact, there may be half of each type of star. These two types of stars have exactly the same spectra. Present astronomical means cannot distinguish between them."*

Dirac believed that the universe should be symmetric. Viewed on a large scale, it should contain one half electrons, one half positrons, one half protons, one half antiprotons, or, in short, one half matter and one half antimatter, the two being numerically equal.

However, facts have shown that Dirac may have been too partial in the symmetry he had discovered, for the real universe is not symmetric.

The Amount of Antimatter

Just as Dirac pointed out, even if there were stars made of positrons and antiprotons, you would not be able to identify them by spectral means. Between the various atoms and ions, and their corresponding anti-atoms and anti-ions, there is no difference in the optical properties. However, modern astronomy already has the means to determine whether one half of the universe is antimatter.

From the existence of the solar wind, we can state categorically that the whole solar system is made of matter. The solar wind consists of particle streams of protons and helium nuclei, emitted from the Sun and sweeping through the entire solar system. If antimatter is present in the system, then it will meet the solar wind and annihilation will take place. This process will be accompanied by very strong γ-ray radiation. But we have never seen such strong radiation in the solar system. For example, if Jupiter were made of antimatter, then the γ-rays produced by its annihilation with the solar wind that sweeps past would be 10^6–10^8 times stronger than is observed; hence, Jupiter cannot be antimatter. The total amount of antimatter in the solar system should be far less than that of matter.

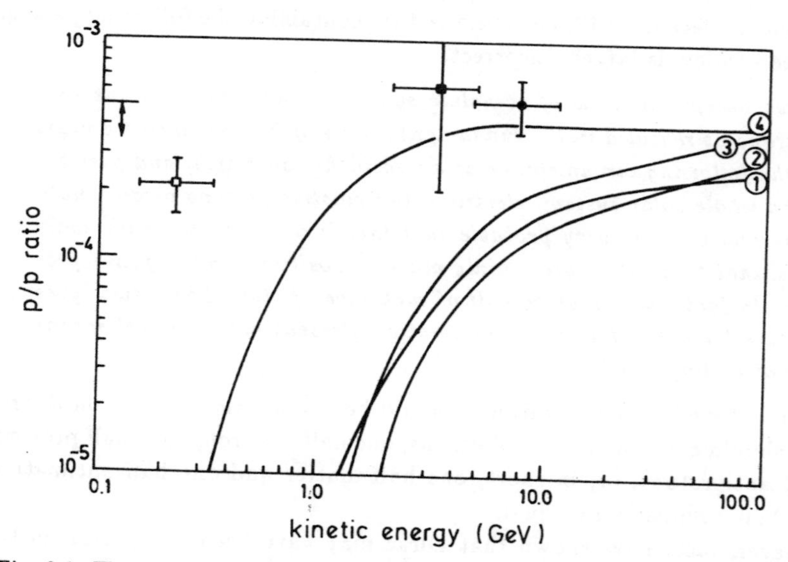

Fig. 9.1. The ratio of the numbers of antiproton to proton observed from cosmic rays.

The situation outside the solar system can be assessed from cosmic rays. The sources of cosmic rays are highly active bodies like pulsars and supernovae. On their passage from their source to the Earth, the cosmic rays only go through extremely tenuous intergalactic medium which will not alter their composition to any extent. Hence the composition of antimatter in the cosmic rays is a reflection of the situation in the source. Figure 9.1 shows the ratio between antiprotons ($\bar{p}$) and protons (p) in the cosmic rays at different energies. We note that the $\bar{p}/p$ ratio is always less than 10^{-4}. By this method, we reckon that the content of antimatter in the entire Milky Way system cannot exceed one-hundredth that of matter.

Over still larger scales, for example, when considering the matter composition of the clusters of galaxies, we can resort to X-ray astronomy. Rather strong X-ray emissions have been observed in many clusters of galaxies. This shows that these objects are filled throughout with hot plasma. The plasma would act the same way in the clusters as the solar wind in the solar system, that is, if the clusters had an equal amount of matter and antimatter, there would be strong annihilation emission due to the presence of the intergalactic gas. However, γ-ray observations have not confirmed any strong γ-ray emissions from clusters of galaxies. Hence, over the range of clusters of galaxies, the content of antimatter does not exceed one-hundredth that of matter either.

For people engaged in γ-ray observations, their results for the solar system and clusters of galaxies have always been "no strong γ-rays observed", which may be disappointing. However, perhaps this is precisely what they had hoped for as the original aim of some of the US observational satellites of γ-rays was to monitor surface nuclear tests by the Soviet Union and other countries. If strong γ-ray emissions can come from celestial objects, this would complicate their monitoring work.

For even larger spatial ranges than clusters of galaxies, we can use the radio emission of distant radio sources to assess the ratio of matter and antimatter. Emissions from radio sources are often polarized. When polarized radiation passes through a region containing a magnetic field, the plane of polarization will undergo rotation called the *Faraday rotation*, or the *Faraday effect.* The direction of rotation depends on whether the region contains mainly electrons or positrons. Since matter and antimatter produce Faraday rotations in the opposite sense, it follows that, if the universe has approximately equal amounts of matter and antimatter, then there will be little net Faraday rotation because the effects caused by the two types of regions will cancel each other out. Yet, nearly every radio source shows Faraday rotation. This proves that the amount of matter and antimatter must be unequal.

In summary, all the evidence supports the assertion of asymmetry between matter and antimatter. Particles and antiparticles are not equal in number, the former being far more numerous than the latter.

The Baryon, Antibaryon and Photon Numbers

The above conclusion will now be written in a more quantitative form.

The conclusion was that galaxies are mainly made of particles, and not antiparticles. Hence, ρ_G is mainly contributed by the rest mass of neutrons and protons. It was stated in Chapter 5 that $\rho_G \sim 10^{-31}$ g/cm^3. Then, with the rest mass of neutrons and protons, $m_n \sim m_p \sim 10^{-24}$ g, we find the number density of baryons (including neutrons, protons, etc.) in the universe to be

$$n_B = \frac{\rho_G}{m_n} \sim 10^{-7}\text{cm}^{-3} . \tag{9.1}$$

On the other hand, there are very few antibaryons.

$$(n_{\bar{B}})_0 \sim 0 . \tag{9.2}$$

For comparison, let us estimate the number density of photons. It was calculated in Chapter 7 that the mass density of the cosmic background radiation is

$$\rho_r = 4.47 \times 10^{-34}\text{g/cm}^3 . \tag{9.3}$$

In the black body radiation at temperature T, the average mass of each photon is about kT/c^2. Hence the number density of photons should be

$$n_r = \rho_r/(kT/c^2) \sim 10^3\text{cm}^{-3} . \tag{9.4}$$

Here we have used $T \simeq 2.7$K. From (9.1) and (9.4) we find the numerical ratio between photons and baryons to be

$$(n_r/n_B)_0 \sim 10^{10} . \tag{9.5}$$

The numbers given at (9.1)–(9.3) and (9.5) can be regarded as the present-day abundances of the particles.

Asymmetry of 10^{-10}

Why is it that, in the basic laws of Nature, particles and antiparticles are completely symmetric, whereas in actual existence the two are so very asymmetric? According to Dirac's symmetry conjecture, we should have

$$n_{\bar{B}}/n_B = 1 , \tag{9.6}$$

whereas, in reality, we have

$$(n_{\bar{B}}/n_B)_0 \sim 0 \; . \tag{9.7}$$

Why is this so?

The simplest answer is that the amount of particles and antiparticles in the universe was asymmetric right from the beginning, and that is why they are asymmetric today. This answer is at least aesthetically unsatisfying. Nonetheless, let us estimate what degree of asymmetry must have existed in the early universe to produce the present asymmetry (9.7).

When the cosmic temperature was greater than 10^{13}K, the condition $kT > m_n c^2$ was satisfied. According to the criteria discussed in Chapter 7, there should have been, at that time, many protons and antiprotons, and neutrons and antineutrons. Moreover, each species would have about the same number density as the photons,

$$n_B \sim n_{\bar{B}} \sim n_r \; . \tag{9.8}$$

If, at that epoch, the number of baryons and antibaryons were precisely equal,

$$n_{\bar{B}}/n_B = 1 \; , \tag{9.9}$$

then, after the cosmic temperature had fallen below 10^{13}K, baryons and antibaryons would undergo annihilation and cancel each other one for one, so there would be no baryons or antibaryons. This is obviously wrong.

Hence, the initial condition of the universe cannot be (9.9); it should be

$$n_b > n_{\bar{B}} \; . \tag{9.10}$$

This way, after annihilation, some baryons would still be left.

Since annihilation reduces the number of baryons and antibaryons, they are not separately conserved during the expansion from $T > 10^{13}$K to $T < 10^{13}$K. Considering the feature of baryons and antibaryons annihilating in pairs, the difference between baryon number and antibaryon number should be conserved. According to Chapter 7, this conservation may be expressed as

$$n_B - n_{\bar{B}} \propto R^{-3} \; . \tag{9.11}$$

Besides, for the photon number n_r defined at (7.2) we also have

$$n_r \propto T^3 \propto R^{-3} \; . \tag{9.12}$$

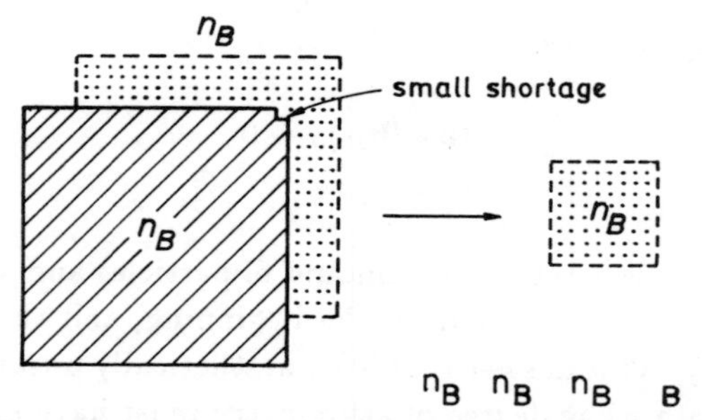

Fig. 9.2. The expansion of the universe can magnify a very tiny original asymmetry into a very enormous asymmetry today.

Hence the quantity defined by

$$S = n_r/(n_B - n_{\bar{B}}) \tag{9.13}$$

does not vary with the scale factor R. It is a conserved quantity during the expansion or

$$S(\text{today}) = S(T > 10^{13}\text{K}) \ . \tag{9.14}$$

From (9.1) and (9.4) of the previous section, we have

$$\begin{aligned} S(\text{today}) &= [n_r/(n_B - n_{\bar{B}})]_0 \\ &= (n_r/n_B)_0 \\ &\sim 10^{10} \ . \end{aligned} \tag{9.15}$$

Also, from (9.8), we have

$$\begin{aligned} S(T > 10^{13}\text{K}) &= n_r/(n_B - n_{\bar{B}}) \\ &\sim n_B/(n_B - n_{\bar{B}}) \\ &= 1/[1 - (n_{\bar{B}}/n_B)] \end{aligned} \tag{9.16}$$

substituting (9.15) and (9.16) into (9.16) we have

$$\begin{aligned} n_{\bar{B}}/n_B &= 1 - 10^{-10} \\ &= 0.999\,999\,999\,9 \ . \end{aligned} \tag{9.17}$$

In other words, in order to produce today's extreme asymmetry between baryons and antibaryons, we need only a minute asymmetry of $\sim 10^{-10}$ at the beginning of the universe, or the time when $T > 10^{13}$K. The expansion of the universe will magnify the tiny original asymmetry into the enormous asymmetry of today.

This story of asymmetry is seemingly quite perfect.

The Story of Gong Gong

The story of asymmetry is still not thorough enough when judged from a strict, evolutionary point of view. True, 10^{-10} is very tiny, but it is still asymmetry. An out-and-out evolutionary stand would hold that everything starts from simplicity, symmetry, and equilibrium; all complex, asymmetric, unbalanced phenomena are not so initially. They all had an origin.

Accordingly, asymmetry must have an origin in symmetry. The Astronomical Chapter ("Tian Wen Xun") of "Master Huai Nan" ("Huai Nan Zi"), records a famous legend:

> *"Once upon a time, Gong Gong was contending the throne with Zhuan Xu. He got angry and touched Mt. Bu Zhou. The Heavenly pillar snapped and the Earthly mainstay broke. The Heaven leaned to the northwest, thither the Sun, the Moon and the stars moved; the Earth was no longer filled in the southeast, thither water and dust collected."*

Judged by cosmological standards, this story seems to state that all asymmetries of the universe came from an originally symmetrical universe. The original Heaven and Earth was level and not inclined. The fight between Gong Gong and Zhuan Xu changed it into the present, observed "inclined Heaven" and "unfilled Earth".

According to this viewpoint, we should explain how to form the 10^{-10} asymmetry from symmetry in the early universe. Who in modern cosmology can play the role of Gong Gong, the Destroyer?

The Destroyer represents the process of non-conservation of baryons.

In particle physics, there is a conservation law called the conservation of the baryon number, meaning that in any particle process, the difference (baryon number minus antibaryon number) is conserved. It is usually abbreviated as "B-conservation". Obviously, if the B-conservation really holds, then an $n_B = n_{\bar{B}}$ state can never evolve into an $n_B \neq n_{\bar{B}}$ state; even if the two differ by only 10^{-10}, this line still cannot be crossed.

Therefore, only when we have found a process of B-nonconservation will it be possible for us to talk about the universe developing from a symmetric state of matter to an asymmetric state of matter.

The Grand Unified Theory and B-Nonconservation

Not until the Grand Unified Theory (GUT) did we find the basis for the possibility of B-nonconservation.

To give a systematic introduction to GUT would require a book at least as long as the present one. Hence only the simplest and fastest sketch will be given.

The aim of GUT is to attempt to unite strong, weak and electromagnetic interactions.

We are all familiar with the electromagnetic interaction. It is determined by the electric charge, and it acts only between electrically charged particles. Electromagnetic interaction is due to the emission and absorption of photons by charged particles.

Similarly, the strong interaction is determined by the so-called color charge, and it is due to the exchange of gluons between the color-charged particles. There is only one kind of electric charge for the electromagnetic interaction, but the color charge for the strong interaction has three kinds, called red, green and blue color charges respectively.

All particles participating in the strong interaction, like baryons and some mesons, are made of quarks having three "colors": red, green and blue. Each colored quark can have one of six different "flavors": up, down, strange, charm, bottom and top. Hence, there are altogether $3\times6 = 18$ different sorts of quarks, and of course, each quark has its own antiquark. These properties of quarks are listed in Table 9.1.

Table 9.1 Colors, flavors and electric charges of quarks.

Flavor	Electric charge	Color
up (u)	$\frac{2}{3}$e	red, green, blue
down (d)	$-\frac{1}{3}$e	red, green, blue
strange (s)	$-\frac{1}{3}$e	red, green, blue
charm (c)	$\frac{2}{3}$e	red, green, blue
bottom (b)	$-\frac{1}{3}$e	red, green, blue
top (t)	$\frac{2}{3}$e	red, green, blue

Different colors may cancel each other out. For example, a red quark, a green quark and a blue quark together form a "colorless" system. A quark and an antiquark also form a colorless system. Such colorless systems do not interact with gluons, just as electrically neutral bodies do not participate in the electromagnetic interaction.

All naturally occurring systems in Nature are colorless. All baryons are made of three quarks, one of each color. For example, a neutron is $n = ddu$, a proton is $p = uud$. Each meson is made of one quark and one antiquark and is also colorless. For example, $\pi^- = d\bar{u}$, $\pi^+ = \bar{d}u$, etc.

The weak interaction can also be described in similar terms. It is determined by two other kinds of charges, and is due to the exchange of the W^+, W^- and Z^0 particles. These particles have been discovered in 1983. Their discovery verifies the theory.

Six particles participate in the weak interaction and not in the strong interaction. They are the electron (e), electron neutrino (ν_e), muon (μ), mu neutrino (ν_μ), tauon (τ) and tau neutrino, (ν_τ). They are called leptons. The 18 quarks and the 6 leptons are the basic elements that make up all matter in the universe. Table 9.2 lists some groups of basic elements formed of quarks and leptons. The subscripts R, G, B denote red, green, and blue, respectively.

Table 9.2 Basic elements of matter.

	Leptons	Quarks		
First Generation	e^-	d_R	d_G	d_B
	ν_e	u_R	u_G	u_B
Second Generation	μ^-	s_R	s_G	s_B
	ν_μ	c_R	c_G	c_B
Third Generation	τ^-	b_R	b_G	b_B
	ν_τ	t_R	t_G	t_B

Quarks also participate in the weak interaction, having weak interaction colors. Hence baryons and mesons also have weak interaction colors. The neutron decay is just one kind of weak interaction,

$$n \to p + e^- + \bar{\nu}_e \,.$$

In terms of quarks, the neutron decay is

$$\begin{array}{rcccc} & d & \to u + W^- \to & u + e^- + \bar{\nu}_e \\ & d & \to & d \\ + & u & \to & u \\ \hline & ddu & \to & uud + e^- + \bar{\nu}_e \end{array}$$

It is induced by the W^- particle. The $W^\pm$ and Z^0 particles are different from photons or gluons, which have zero rest mass. The W^-, $W^\pm$ and Z^0 particles however have very large masses. Hence the weak interaction is a force with a very short range.

In summary, the strong, weak and electromagnetic interactions have a total of six different charges. To unite the three interactions requires some new process which can change the color charges of the strong interaction into those of the weak interaction. An important result of such a process is to make baryons capable of becoming mesons and leptons, thus destroying the B-conservation.

Suppose a certain X particle can induce a color charge of the strong interaction to change into a weak interaction, it may give rise to the following typical process of B-nonconservation:

$$\begin{array}{rcccl} & u & \rightarrow \bar{u} + X \rightarrow & \bar{u} + \bar{d} + e^+ \\ & u & \rightarrow & u \\ + & d & \rightarrow & d \\ \hline & uud & \rightarrow & d\bar{u} + \bar{d}u + e^+ \end{array} .$$

Since $p = uud$, $\pi^- = d\bar{u}$, $\pi^+ = \bar{d}u$, the above process is just proton decay

$$p \rightarrow \pi^- + \pi^+ + e^+ .$$

Proton decay is one of the most important instances of B-nonconservation. Experiments regarding this have already been discussed in Chapter 1. The present estimate gives a lower limit to the lifetime of 10^{31} years; hence the mass of the X particle must be very large, otherwise the proton life would be too short. Some Grand Unified Theories estimate the mass to be 10^{15} times the proton mass, or 10^{15} GeV.

Baryon Asymmetry

Since the mass of the X particle is very large, the B-nonconservation process such as proton decay in the present universe only has a very small effect on the properties of the universe.

From the chronological table of the thermal history of the universe, given in Chapter 7, we see that when the age of the universe was less than 10^{-36} seconds, the temperature exceeded 10^{28}K, and the energy scale was greater than 10^{15} GeV. Many X particles were present, together with their antiparticles, $\bar{X}$. B-nonconservation events were very common then. However, because the entire state was one of thermal equilibrium, such events would not lead to asymmetry between baryons and antibaryons. Their numbers would be kept equal.

Once the age of the universe reached 10^{-36} seconds, the situation changed. The energy scale of the universe began to fall below 10^{15} GeV; X and $\bar{\text{X}}$ also began to disappear through annihilation or decay.

If the X-product can lead to proton decay, then it must possess the following two modes of decay:

$$\begin{aligned} \text{X} &\to q + q\,, \quad r \\ \text{X} &\to \bar{q} + l\,, \quad 1 - r \end{aligned} \tag{9.18a}$$

X can decay into either two quarks, with probability r, or one antiquark and one lepton, with probability $1 - r$. For the antiparticle $\bar{\text{X}}$, we have

$$\begin{aligned} \bar{\text{X}} &\to \bar{q} + \bar{q}\,, \quad \bar{r} \\ \bar{\text{X}} &\to q + \bar{l}\,, \quad 1 - \bar{r} \end{aligned} \tag{9.18b}$$

expressing the circumstance that $\bar{X}$ can decay into either two antiquarks, with probability $\bar{r}$, or one quark and one antilepton with probability $1 - \bar{r}$.

If $r \neq \bar{r}$, then after the decay of one X and one $\bar{\text{X}}$, the numbers of quarks and antiquarks will, on the average, be different. The difference will be

$$2r + (1 - \bar{r}) - (1 - r) - 2\bar{r} = 3(r - \bar{r})\ .$$

Since a baryon is formed of 3 quarks, the B-number of each quark is 1/3. Hence, after the decay of a pair of X and $\bar{\text{X}}$, the B-number will be increased by

$$\varepsilon = r - \bar{r}\ , \tag{9.19}$$

and this will lead to asymmetry between baryons and antibaryons.

Each $X\bar{X}$ pair gives a B-number asymmetry of size ε and each unit volume has $n_X \sim n_{\bar{X}}$ pairs. Hence, after all have decayed, each unit volume will have contributed to the B-number,

$$n_B - n_{\bar{B}} \sim n_X \varepsilon\ .$$

At the age of 10^{-36} seconds, $n_X \sim n_B$, and the above can be re-written as

$$n_{\bar{B}}/n_B \sim 1 - \varepsilon\ . \tag{9.20}$$

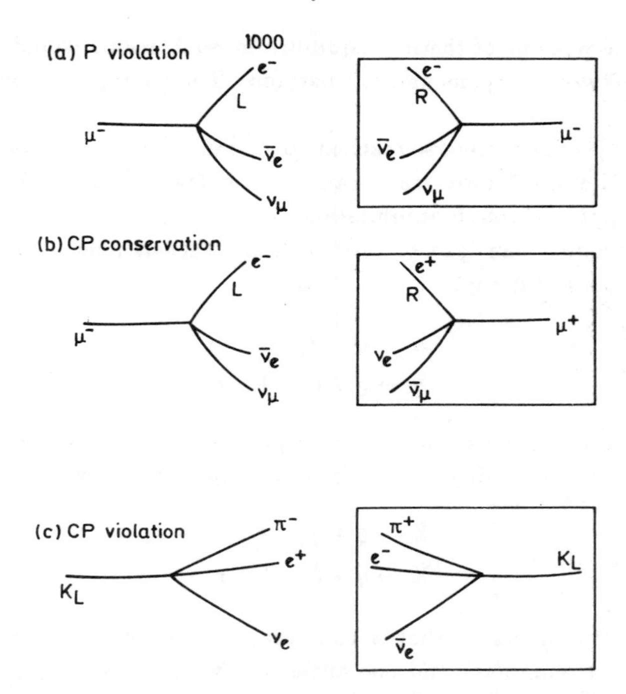

Fig. 9.3. Several decay processes in particle physics. L and R denote the left-handed and right-handed rotation respectively. (a) The half-life period of μ decay is much shorter (1 to 1000) than that of its mirror process shown in the left frame. This is parity (P) violation. (b) The half life period of μ decay is the same as its mirror plus the anti-particle process shown in the left frame. This is the conservation of charge conjugation plus parity (CP); (c) The lifetime of K_L decay is different from that of its mirror plus anti-particle process shown in the right frame. This is the violation of CP. K_L is the anti-particle itself.

Comparing this with (9.17), we see that we need only have

$$\varepsilon \sim 10^{-10} \tag{9.21}$$

in order to explain the baryon asymmetry in the universe.

Seen from the standpoint of particle physics, the condition (9.21) is equivalent to saying that the decay processes of X and $\bar{X}$, (9.18a) and (9.18b), are C and CP-violating. The violation is very small, being only about 10^{-10}.

Weak CP-violation processes are no strangers in particle physics. A very famous example is the decay of long-lived, neutral K_L mesons. These mesons decay in the following two modes:

$$K_L \to \pi^- + e^+ + \nu_e \ ,$$
$$K_L \to \pi^+ + e^- + \bar{\nu}_e \ .$$

Since K_L is its own antiparticle, if the CP symmetry is maintained, then the two modes must have the same probability of occurrence, that is, K_L is as likely to decay into a negative π-meson as into a positive π-meson. But experiments showed that the two decay modes are not symmetric, and that the decay mainly proceeds in the first mode, that is, CP is violated. (For details, see Fig. 9.3).

The CP violation in K_L is extremely weak. It is not of any importance in today's universe either. But if the X decay has a very weak CP violation, then this will lead to a minute asymmetry between baryons and antibaryons in the earliest universe, which will then explain why there is such a large baryon/antibaryon asymmetry today.

In short, if such X-particles conjectured in the Grand Unified Theory do exist, then the matter/antimatter asymmetry will naturally arise in the course of cosmic expansion.

A sufficiently quantitative theory for the various aspects of the X-particle has not yet been worked out. Hence the above theory of the generation of asymmetry is still quite imperfect. Nevertheless, the theory has had good appraisals because it has very harmoniously combined the unification in particle physics and the generation of matter in cosmic evolution. Furthermore this theory emphasizes that the asymmetry of today is a result of the breaking of symmetry, and this reminds us of a truth in aesthetics,

$$\text{symmetry} + \text{defect} = \text{beauty} .$$

Indeed, one can find many instances illustrating this truth in the arts: gardens in studied randomness, Venus with arms broken off, the Qin funeral warriors and horses in strict, symmetrical formation but all with different expressions. Why does symmetry plus "defect" equal beauty? Perhaps the universe evolved in the form of symmetry plus defect.

Chapter Ten

INFLATION OF VACUUM

The Non-Unique Vacuum State

What is vacuum? People at different times have understood vacuum differently.

In the time of classical mechanics, vacuum was simply "empty space"; in the time of Maxwell's electromagnetism, vacuum was an aether; in modern physics, vacuum is the base state — the starting point of various excitations. Although the understandings are different, they all tacitly agree that vacuum is unique and that only one determinate vacuum state exists. The vacuum state is the starting point and basis that provides measurements for all motions, and at the same time is itself free from any effect of the motions of matter. The relation between the two is one-way, namely

$$\boxed{\text{vacuum}} \longrightarrow \boxed{\text{motion of matter}}\ .$$

Such a unidirectional relationship is ill-fitted for the usual framework in physics, where the effect between systems is always mutual in character.

In the modern theory of particle physics, we are beginning to recognize that there is indeed the type of mutual relationship that we expect between the vacuum and material motion, that is,

$$\boxed{\text{vacuum}} \rightleftharpoons \boxed{\text{state of matter}}\ .$$

This implies, above all, that the vacuum state is certainly not unique, and that there are many possible vacuum states.

What do we mean by "many different vacuums"? Let us take the simplest example. Suppose a field exists in Nature, with its intensity denoted by σ. Generally "empty space" should correspond to $\sigma = 0$. At the vacuum state, the field intensity is zero. Hence, from the viewpoint of particle theory, vacuum corresponding to $\sigma = 0$ is just due to $\sigma = 0$ being the lowest energy state; all $\sigma \neq 0$ states have higher energies. Figure 10.1 shows the variation of the field energy with the field intensity. The point $\sigma = 0$ is a minimum, hence it corresponds to the base state or the vacuum state.

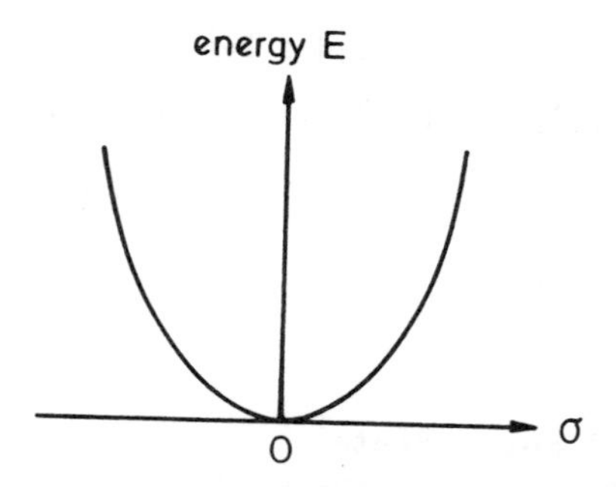

Fig. 10.1. Energy of σ field as a function of field intensity σ. $\sigma=0$ is the minimum, namely, the state of vacuum.

According to the above interpretation, vacuum does not necessarily correspond to no σ-field or the field at $\sigma = 0$. If the energy of the σ-field is of the form shown in Fig. 10.2, then there will be two vacuum states,

$$\sigma = +\sigma_0\,, \quad \sigma = -\sigma_0\,. \tag{10.1}$$

For this energy curve, although $\sigma = 0$ is an extreme value, it is a maximum, so the $\sigma = 0$ state is unstable.

The Grand Unified Theory believes that there is, in Nature, the so-called "Higgs fields", for which the energy curve is indeed of the form shown in Fig. 10.2. In the vacuum state, the Higgs field σ is not zero.

Again, from the energy curve of Fig. 10.2, we see that the $\sigma = 0$ state is comparatively symmetric, while for the $\sigma = +\sigma_0$ or $\sigma = -\sigma_0$ states the symmetry is broken. A basic viewpoint of the Grand Unified Theory is this: the reason why Nature cannot be kept in a state of perfect symmetry is that the completely symmetrical vacuum state is unstable. First, vacuum loses complete symmetry, and this then leads to asymmetries in other states or in physical processes. This mechanism is called the *spontaneous breaking* in the symmetry of vacuum.

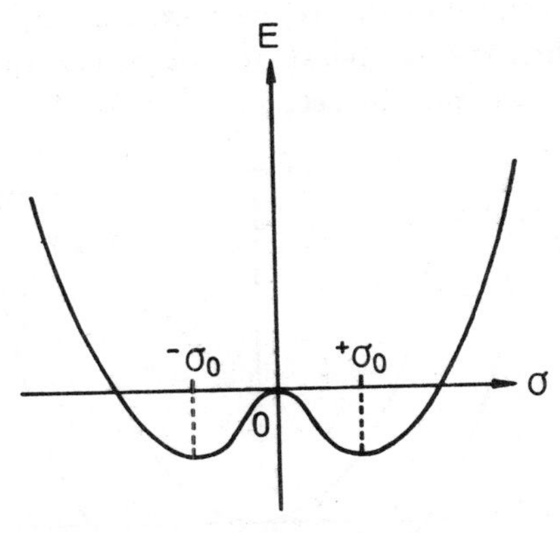

Fig. 10.2. Energy of σ field as a function of field intensity σ. Now the states of vacuum are $\sigma=+\sigma$ or $\sigma=-\sigma_0$. $\sigma=0$ is a maximum. It is unstable.

At this point, we are faced with the following problem. In the last chapter, we emphasized that the asymmetry in the universe today evolved from a completely symmetric, early stage. Here we say that the basic asymmetry in Nature comes from symmetry breaking out of the vacuum state. Do these two points of view contradict each other? Can they be unified?

Yes, they can be unified. The way to unify them is in the *phase transition* of the vacuum of the universe.

Phase Transition of Vacuum

Phase transition is no stranger. For example, when water is cooled to 0°C, a change from the liquid phase to the solid phase will occur. The water before and after the phase change has the same chemical composition, but the symmetry of its state has changed. Again, ferromagnetic materials possess ferromagnetism at temperatures below the Curie temperature and lose this property above. This is also a phase transition; before and after the transition, properties of symmetry are altered.

The characteristic feature of the phase change of vacuum is also a change in properties of symmetry. Let us again use the above σ-field as an example. Strictly speaking, the curve of Fig. 10.2 refers to matter at temperature $T = 0$. When $T \neq 0$, the curve should be as shown in Fig. 10.3. Here the features of the curves are as follows:

When $T < T_c$, a minimum occurs at $\sigma \neq 0$;

When $T > T_c$, a minimum occurs at $\sigma = 0$.

Here T_c is the critical temperature. When the temperature of matter is higher than the critical temperature, vacuum is safely located in the symmetric $\sigma = 0$ state; when the temperature falls below the critical temperature, the symmetry of the vacuum is lost.

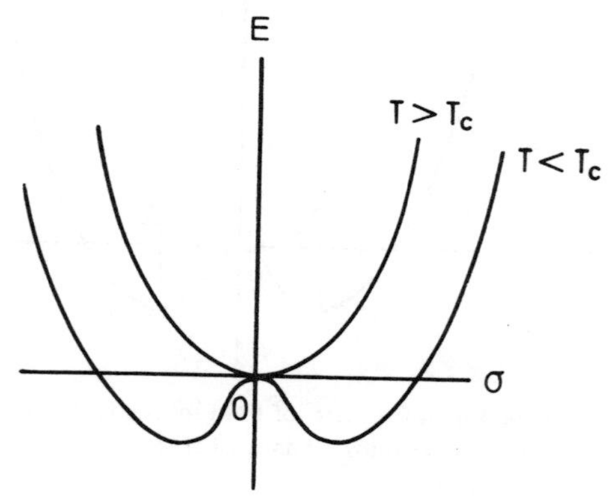

Fig. 10.3. The curves of the σ field energy are dependent on temperature.

The thermal history of the universe is one of decreasing temperature. Hence in the course of cosmic evolution, a series of vacuum phase changes must have occurred. Moreover, these changes always occur from the symmetric towards the less symmetric. Many asymmetries of today's universe have evolved from vacuum phase changes.

The most marked asymmetry of today's universe is the existence of four interactions of vastly different strengths: strong, electromagnetic, weak and gravitational interactions. This agrees with the viewpoint that everything was formed during the evolution of the universe. During the earliest stage of the universe, when the highest degree of symmetry existed, these four interactions were indistinguishable. Only one "variety" was present. This could be described using the *superunified theory*. As the universe cooled off, successive vacuum phase changes corresponding to supergravity, Grand Unification, and electroweak unification occurred (see Fig. 10.4). Each phase change caused a lowering of the symmetry and the differentiation of a particular interaction. The present four interactions have not always been such from the beginning; they were generated in stages.

The epoch at which a phase change occurs is determined by the critical temperature. The unified electroweak theory is well matured; its critical temperature is $kT_c \sim 100$ GeV. The Grand Unified Theory has also had some success; its critical temperature is $kT_c \sim 10^{15}$ GeV. As for super unification,

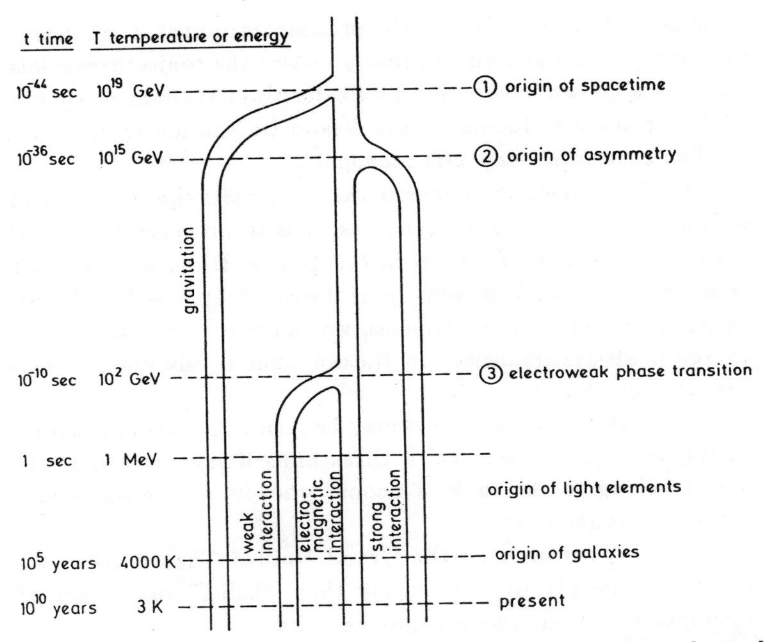

Fig. 10.4. In the earliest universe, none of the four interactions that are presently observed existed, only one unified one. As the universe expanded and cooled off, successive vacuum phase transitions took place, such as phase transitions of supergravity at 10^{-44}s, Grand Unification at 10^{-36}s, electroweak unification at 10^{-10}s. Every phase transition lead to a decrease in symmetry. The present interactions were also formed during the vacuum phase transitions.

we still lack a theory that can be confirmed with observations. A value that we can use as a temporary measure is $kT_c \sim 10^{19}$ GeV. Thus, according to the chronological table in Chapter 7, the electroweak phase change occurred at cosmic age $t \sim 10^{-12}$ seconds, the Grand Unification phase change at $t \sim 10^{-36}$ seconds, and the super unification possibly at $t \sim 10^{-44}$ seconds.

Fossils of the Grand Unification Phase Change

By what means can we prove that vacuum phase changes must have taken place in the early universe?

Some say that cosmology is a kind of archaeology. This is a reasonable simile. In archaeology we use the relics of pre-historical events to reconstruct the human life at the time. In cosmology we use the "fossils" that were left behind from early cosmic events to confirm or refute theoretical models about the early universe. The helium abundance of ~ 0.30 is a "fossil" of the early event of nucleosynthesis, and the proton/antiproton asymmetry is a "fossil" of the early particle genesis. Similarly, to test the conjectures relating to vacuum phase change, we have to search for fossils of vacuum phase change events.

What are the "fossils" of the Grand Unification vacuum phase change?

The answer is: *magnetic monopoles.*

The classical electromagnetic theory stresses that there are electric charges in Nature, but no magnetic charges, that is, no magnetic monopoles, without saying why this is so. Then in the 1930's, Dirac first pointed out that the existence of magnetic monopoles is theoretically possible. Moreover, given the existence of magnetic monopoles, we can very naturally explain why electric charge is always quantized in Nature, that is, always an integer multiple of the electron charge.

Although Dirac's theory is very beautiful, people in general do not believe in magnetic monopoles. Even Dirac himself says, in a letter in 1981, "up to now I belong to the rank of people who do not believe in the existence of magnetic monopoles".

However, research on the Grand Unified Theory and the early universe has forced people to stop basing their "belief" or "disbelief" in magnetic monopoles on their own sentiments.

In 1974, 't Hooft and Polyakov separately showed that in the Grand Unified Theory, the existence of magnetic monopoles is almost a necessity. This is determined by the mechanism of spontaneous breaking of vacuum symmetry of the Higgs field. After spontaneous symmetry breaking takes place, a number of different vacuum states may occur. In the example of the σ-field, two different vacuum states $\sigma = \pm\sigma_0$ exist. Therefore the vacuum breaking may form individual regions such that, within one region, the vacuum state is the same, while different regions have different vacuum states (see Fig. 10.5).

Locations where the regions meet possess all the intrinsic properties of magnetic monopoles. Such magnetic monopoles of the Grand Unified Theory have a mass of about

$$m \approx 10^{16} \text{ GeV/c}^2 \tag{10.2}$$

that is, about 10^{16} times the proton mass.

Next, using methods of cosmology we can estimate how many such magnetic monopoles exist today. Since these objects are at the intersection of regions of different vacuum states, their number should be equal to the number of such regions. If the Grand Unification phase changed at $t \sim 10^{-36}$ seconds, regions possessing causal connection will have a spatial scale of about

$$ct \sim 10^{-26} \text{ cm} , \tag{10.3}$$

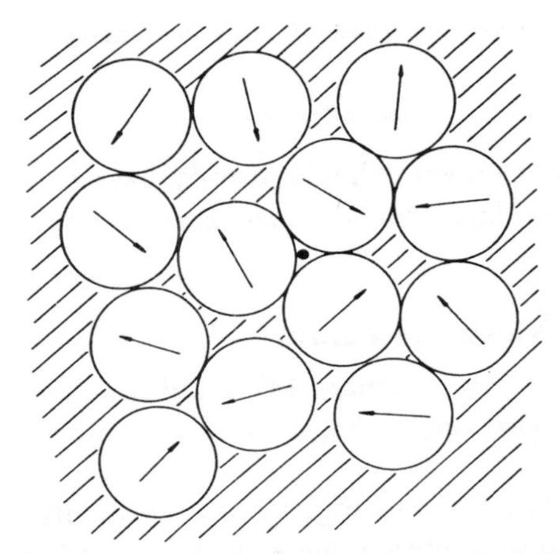

Fig. 10.5. The structure of vacuum may become domain-like due to the phase transition of vacuum. Different domains have different states of vacuum. Some interspaces of the domains are magnetic monopoles.

or, a causally connected volume of

$$V \sim \frac{4\pi}{3}(ct)^3 \sim 7 \times 10^{-77}\ \mathrm{cm}^3\ . \tag{10.4}$$

Exceeding V, there is no causal connection. Hence, after spontaneous symmetry breaking, the size of regions with the same vacuum state cannot exceed V. Thus, at epoch $t \sim 10^{-36}$ seconds, the number density of magnetic monopoles is about

$$n_m \sim \frac{1}{10V} \sim 10^{75}\ \mathrm{cm}^{-3}\ . \tag{10.5}$$

The factor 10 here represents a conservative estimate, that is, we suppose every 10 different vacuum state regions correspond to 1 magnetic monopole.

Annihilations of magnetic monopoles are extremely rare, so their number is conserved in the cosmic expansion. From this we can estimate that their number density today should be:

$$\begin{aligned} n_m &\sim 10^{75}(R/R_0)^3 \\ &\sim 10^{75}(T_0/T)^3 \\ &\sim 10^{75}(2.7/10^{28})^3 \\ &\sim 10^{-8}\mathrm{cm}^{-3}\ . \end{aligned} \tag{10.6}$$

In the above we have used the formulae $n \propto R^{-3}$, $T \propto R^{-1}$ (see Chapter 6), and have taken $T = 10^{28}$K and $T_0 = 2.7$K.

The last chapter gives the present baryon density as $n_B \sim 10^{-6} - 10^{-7}$ cm^{-3}, hence

$$n_m/n_B \sim 10^{-2} , \tag{10.7}$$

that is, every hundred baryons should already have one magnetic monopole. This ratio is much too large, for we must remember that we have not seen a single one yet!

The Event of St. Valentine's Day

On St. Valentine's Day (February 14) in 1982, we seemed to have, at last, received a small consolation, perhaps a magnetic monopole really did fulfill our expectations.

In spring that year, Cabrera of Stanford University announced that his superconducting quantum interferometer had recorded, on Feb. 14, a sudden jump in magnetic flux (Fig. 10.6). This phenomenon can only be interpreted as a magnetic monopole passing through his equipment. Moreover, from the size of the jump, the value of the magnetic charge could be determined, and the result agreed very well with theoretical expectations. Therefore Cabrera believed this to be a magnetic monopole event.

Whether the event of St. Valentine's Day was really caused by a magnetic monopole is not entirely certain. Even so, this affair was widely regarded as significant. Cabrera hoped to see a second similar event, but so far it has been in vain. This means that, even if the event was a magnetic monopole, the number density must be very small. After waiting for hundreds of days, he found only one, hence an upper limit of the abundance is

$$n_m/n_B < 10^{-9} , \tag{10.8}$$

seven orders of magnitude smaller than the predicted value (10.7)!

In fact, Cabrera's upper limit (10.8) is still too high. We can find more stringent limits.

We know that the matter density today is about the critical mass density $\rho_c \sim 10^{-29}\ \text{g/cm}^3$. If all this mass is provided by magnetic monopoles, then their number density should be $n_m \sim \rho_c/m$. The mass of a magnetic monopole is $m \sim 10^{16}\ \text{GeV/c}^2 \sim 10^{-8}$ g, hence $n_m \sim 10^{-21}\ \text{cm}^{-3}$, or an upper limit of

$$n_m/n_B < 10^{-15} . \tag{10.9}$$

We can find an even stronger upper limit in the Milky Way system. Magnetic monopoles, when moving in magnetic fields, will be accelerated, just as electric charges are accelerated in electric fields. Such acceleration will deplete the field energy. Therefore, if too many magnetic monopoles exist, then the magnetic field of the Milky Way would soon be exhausted. From this consideration we can state that the abundance in the Milky Way system is bounded above by

$$n_m/n_B < 10^{-16} . \tag{10.10}$$

Some people are also attempting to find magnetic monopoles directly in certain samples. For example, lunar rocks and meteorites are both old rocks of the solar system and people have thought of using strong magnetic extraction techniques to pull out any monopoles remaining in such samples, and have attempted to find tracks of monopoles through the rocks. All these attempts have failed. No monopoles were found, not even tracks. The upper limit derived this way is

$$n_m/n_B < 10^{-28} . \tag{10.11}$$

All the experimental results seem to show that magnetic monopoles do not exist in reality, or almost do not, whether today, or in the early epoch.

Hence the predicament we now find ourselves in is this: we have not found the fossils that the vacuum phase change in the early universe should have left behind.

Inflation of the Universe

If we do not wish to doubt the conclusion of the Grand Unified Theory regarding the existence of magnetic monopoles, and if we wish to get out of a difficult position, then we must find a way of lowering the production of monopoles in the early universe.

An effective means was found in 1981: *inflation of the universe*. The word "inflation" usually refers to currency, and there is indeed a similarity between the mechanism of the two inflations.

In a free economy the rate of unemployment and the rate of inflation are inversely proportional. If the rate of inflation is suppressed, then the rate of unemployment will rise. A lowering of the rate of unemployment often cannot avoid leading to an increase in inflation. The production rate of magnetic monopoles and the rate of the cosmic expansion are likewise negatively correlated. From the formula (10.5) we see that the number of monopoles is inversely proportional to the size of causal regions at $t \sim 10^{-36}$ seconds, and the latter is determined by the rate of cosmic expansion. Hence, if the expansion is lowered, monopoles will increase in number, and if we wish

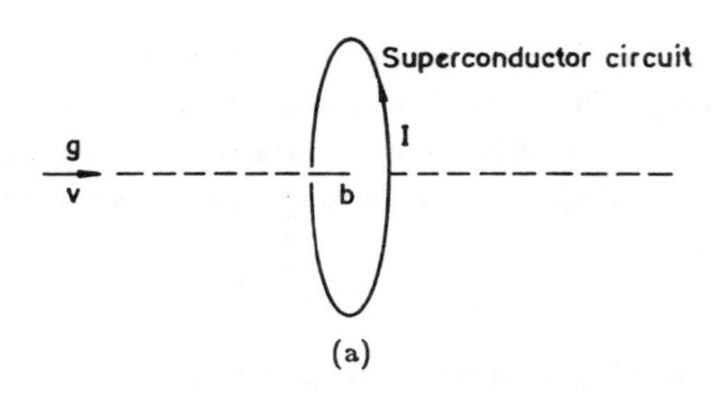

(a)

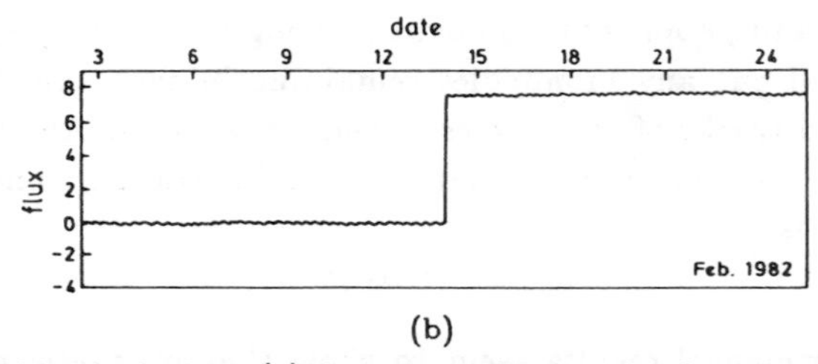

(b)

Fig. 10.6. St. Valentine's day. (a) A superconducting circular wire. When a magnetic monopole crosses the circle, then a current I occurs in the wire. (b) The record of magnetic flux (equal in value of I) given by Cabrera. It shows a jump of the magnetic flux on February 14th, 1982.

to lower the number of monopoles, we must have a high expansion rate of the universe.

According to the inflationary model, when the cosmic temperature drops to T_c or even below, phase changes do not occur and vacuum is still kept in a symmetric state. From the curves of the σ-field given in Fig. 10.7, we see that for $T < T_c$, a potential wall exists between the symmetric states $\sigma = \pm\sigma_0$. Even though the energy at $\sigma = \pm\sigma_0$ is lower, vacuum is still kept at the $\sigma = 0$ state, and a phase change does not take place immediately; thus, a supercooled state is formed, just like water below 0°C is supercooled water.

When the universe is in the supercooled state, both components, particles and radiation, exert little effect on the cosmic expansion. The expansion is mainly determined by the vacuum state itself. The vacuum differs from both radiation and particles in that its equation of state is

$$P = -\rho_v c^2 , \tag{10.12}$$

showing that its pressure P and its mass density ρ_v have opposite signs. If ρ_v is positive, then P is negative.

Roughly speaking, pressure is the density of kinetic energy. Negative pressure corresponds to negative mass. The effect of negative mass is repulsion,

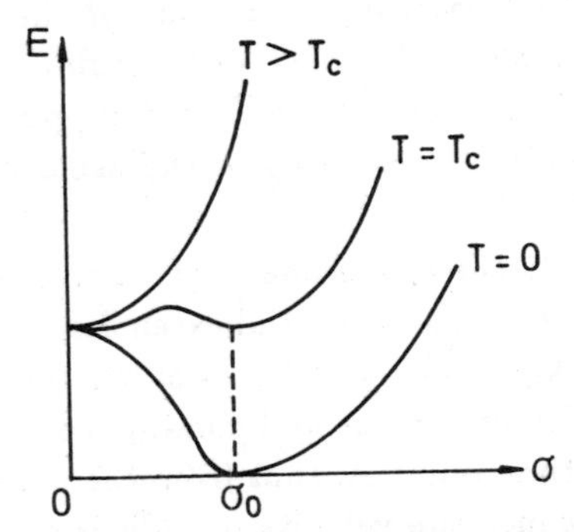

Fig. 10.7. Because there is a potential hill between the states of $\sigma=0$ and $\sigma=\sigma_0$, when the cosmic temperature drops to T_C or even below, phase transition may not be occur. In this case, the universe is overcooling.

not gravitation. Hence, the epoch when a $\rho_v > 0$ vacuum state is dominant is one in which repulsion predominates. Under repulsion, the cosmic expansion does not decelerate. It will lead to extremely fast expansion or, in other words, *inflation*.

Thus, by the time the supercooled phase ends and the universe begins the phase change to form monopoles, the causal volume of the universe may have increased by 10^{30} times the value at (10.4) due to inflation, and the number of monopoles in the whole universe will be practically zero.

The reason therefore that we cannot find magnetic monopoles today is not because they do not exist, but because the universe has never provided the necessary conditions for generating them.

Horizon and Flatness

If the theory of inflation could only explain the excess of magnetic monopoles and nothing else it would be worthless, for, one of the criteria regarding value in physics is *generality*. It is only when a theory can simultaneously explain several different phenomena that it has theoretical significance.

The success of the theory of inflation is due to its ability to solve two other, older problems of the universe:

1. The *horizon* problem. Why is the universe so homogeneous and isotropic?

2. The *flatness* problem. Why is it so hard for us to determine whether the universe is finite or infinite?

Consider first the horizon problem.

In Chapter 2, we took isotropy to be one of the basic assumptions. In Chapter 7 we showed that the observed results on large angular scale anisotropy of the cosmic background radiation support this assumption. The problem is how to explain this isotropy in the usual expanding model of the universe.

Figure 10.8 represents the spacetime of the universe with the vertical axis for time and the horizontal axis for spatial extension. P is our present location in spacetime. The background radiation that we receive today was emitted at time t_R. Radiation directed at us and coming from the positive x-direction originated in the spacetime point A, that from the negative x-direction in the spacetime point B. Because the velocity of light is c, the distance between A and B is very large. Isotropy means that A and B have the same emission properties. In physics, sameness does not come by accident; it is achieved after reaching equilibrium through an exchange of heat, that is, a causal connection should exist between A and B. However, from the beginning of the universe to t_R, the causal range is limited to the region enclosed by a and b. This region is called the horizon, and it is far smaller than the range between A and B. Thus, using ordinary theory, we cannot explain isotropy.

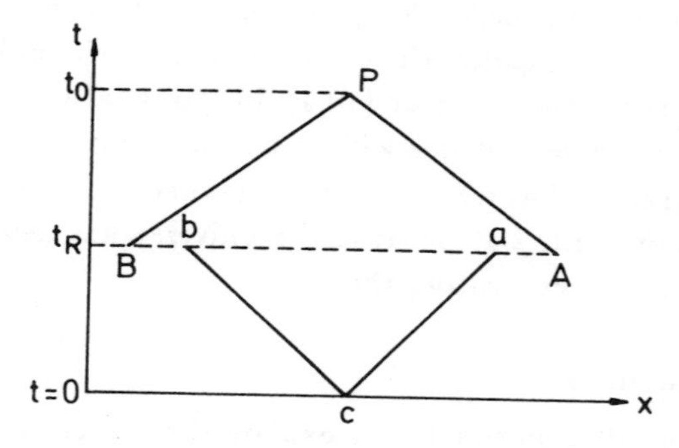

Fig. 10.8. The space-time of the universe. P denotes our position. We can observe the events between A to B at time t_R. However, the causal region at t_R is only limited by the scale of a to b; ab is much smaller than AB.

In the inflationary model, the universe expands so fast that the causal range or horizon can be much greater than ab, and can easily include A and B. Isotropy is then explained in a natural manner.

Now consider the flatness problem.

In Chapter 4, we saw that the long drawn-out debate on whether the universe is finite or infinite has so far not been settled. The reason is that the cosmic mass density ρ_0 is rather close to the critical mass density ρ_c,

$$\rho_0 \sim \rho_c \, . \tag{10.13}$$

Hence, it has been difficult to decide whether $\rho_0 > \rho_c$ (universe finite) or $\rho_0 < \rho_c$ (universe infinite).

Let us put aside for the time being the finite/infinite question, and ask, why is ρ_0 so close to ρ_c? We know that, in the theory of cosmic expansion, ρ_0 and ρ_c are two mutually independent quantities, just as the initial position and initial velocity of an object are mutually independent. Hence, if we assign to them two arbitrary values, then most likely we shall have either $\rho_0 < \rho_c$ or $\rho_0 > \rho_c$ and very rarely $\rho_0 \sim \rho_c$. In order not to have to explain this coincidence in terms of accident, we must look for some mechanism that will adjust ρ_0 to ρ_c.

Inflation can influence such an adjustment. We can prove that, for an expansion determined purely by vacuum, we must have

$$\rho_v = \rho_c \, .$$

This property is also a consequence of the equation of state for vacuum (10.12). Hence, whatever the original circumstances, as soon as inflation is large enough, it will adjust the cosmic expansion to the $\rho_0 \sim \rho_c$ situation.

The Beginning of the Universe

Inflationary cosmology has led us to the beginning of the universe.

The aim of cosmology is to explain the origin of all things. "Tao generates One, One generates Two, Two generates Three, Three generates all things", and cosmology pursues the "One" that is the origin of all things.

The "One" seems to be just the epoch of cosmic inflation. At this epoch, there were no stars, no chemical elements, no particles and no radiation. Only vacuum was in complete symmetry in the expanding universe. Where can we find a simpler, a more symmetric state than that?

The spectacular universe we see today gradually grew out of this very simple, very symmetric "One". The entire evolutionary picture can be sketched as follows: first, at the termination of inflation, a phase change took place in the vacuum, and symmetry passed into asymmetry. This is the earliest origin of asymmetry. The energy released in the phase change became radiation and particles, and this is the origin of matter in the universe. Next, the particle/antiparticle asymmetry was generated, producing the baryons seen today. Then, nucleosynthesis saw the creation of helium and deuterium, and this marked the beginning of chemical elements. Lastly, gravitational assembling gave rise to various stars. Life appeared through evolution, and then the human being, up to today's society, today's world.

At this point, the picture of cosmic generation seems complete, at least in outline. However, cosmologists do not regard it as complete enough and ask: that very simple, very symmetric "One", what did *it* evolve from? How were time, space and vacuum, which we take to be the beginning of the universe, created?

Chapter Eleven

PHYSICS OF THE FIRST MOVE

The First Mover

It is curious that some philosophers who say they do not believe in religion or theology should maintain that the "First Mover" is God. In fact, this assertion itself is pure theological dogma. It comes from Thomas Aquinas, the same author who wrote "Science is a servant of Theology".

In his "Theological Compendium" Ser. 1, Pt. 1, Aquinas wrote:

> *"In the world, certain things are in motion. This is clear and sure from our sensation. Whenever a thing moves, it is being pushed by some other thing. If a thing itself is moving, and is necessarily being pushed by some other thing, then this other thing must be pushed by something else. But we certainly cannot extend this argument one by one till infinity. Hence, it is necessary that eventually we shall arrive at a First Mover who is not moved by anything else. Everybody knows this First Mover is God".*

This is the first edition of the theory of the First Mover in theology. It appeared in the thirteen century.

However, the "First Mover" was not a theological original; quite the opposite, it was already discussed in detail in Aristotle's "Physics". In his discussion of the cause of movements of bodies, Aristotle proved that if we admit that "anything moved is moved by something", then we must necessarily have "a First Mover that is not moved by any other thing".

We see therefore that historically, the first move first appeared as a physical problem, and it was only later that it was used by Aquinas as a method of proving the existence of God. Obviously we must not, because of the theological use, equate the first move with God, even less disqualify it as a physical problem.

In fact, in today's physics, the problem of the first move still exists. Take Newtonian mechanics as an example. In order to explain or predict the motion of objects, we either need to know or fix the initial conditions. The "give" of the initial conditions is equivalent to Aristotle's "First Mover". When we trace back the origin of motion, we shall certainly be led to ask, on what basis are the earliest initial conditions given? This is the problem of the first move.

Physics after Newton has made a number of radical revisions of Newtonian mechanics, but the problem of initial conditions is always there. The basic framework of dynamics remains this: in order to explain the motion of a system, we must either know or fix the initial conditions of the system. In the search for the origin of motion, we still cannot avoid the necessity of needing to know the initial conditions of the universe.

Why does the universe choose this and not that initial condition? If we cannot answer this question, then we are tacitly admitting that physics can only explain the world in the following way: things are so because the first move was like that. Obviously, this is only one half of science.

The Predicament of Infinity

In order to "solve" the other half, classical physics has assumed that the universe is infinite in time, that is, the universe has had an infinite past. If the causal chain of things is infinitely long, then a "first" move will not exist.

This thesis of infinity is not so much a solution of the problem as a temporary measure that evades the problem. In fact, firstly, there is no scientific evidence that the universe had an infinite past, and secondly, there is no scientific evidence to show that infinite time can always exclude the first move. On the contrary, modern cosmology has uncovered more and more evidence to support the view that the universe is finite in time — we have already discussed this in detail in Chapter 3. Also, it has gradually become clear that it is not at all certain that infinite time can avoid a first move.

In fact, Aristotle already thought of using infinite time to avoid the first move, but at the same time he doubted whether infinity could really do this. He said:

"There are difficulties regarding infinite theories. If we deny infinity, we obviously come to an impasse in many places. If we admit infinity, then we still have to answer the following questions. Is it to be regarded as a substance? Or an intrinsic property of something? Or neither? The special task for natural philosophers is this: to make a thorough enquiry as to whether there is some quantity which we can feel to be infinite".

The infinite thesis supposes that time is this sensibly infinite quantity. Thus, in order to exclude the first move, the key point is to prove that there is no existence beyond infinite time. But it is impossible to complete this proof.

We already stated in Chapter 4 that the finite is not necessarily the bounded, and that the infinite is not necessarily the unbounded. For example, black hole physics tells us that infinite time can have a boundary.*

In short, it is not certain that infinite time can avoid the problem of the first move.

Big Bang's Hell

For modern cosmology, the first move is still a difficult problem.

The whole of modern cosmology, from the discovery of cosmic expansion to the proposal of the theory of inflation, has been something of a success. Especially after the appearance of the inflation theory, we now have a fairly complete picture of evolution. It tells us how the universe started from the earliest vacuum, which has almost nothing, to form Heaven, Earth and today's world.

Because of its success, this set of theories has been called standard cosmology. It is popularly known as *the big-bang theory*.

For standard cosmology, the first move is no longer a problem in principle, based on rational thought, but a physical problem. Now, the problem of the first move can be put more specifically: the time, space and vacuum at the epoch of inflation, how did they come about?

Using only the classical big-bang theory, we cannot answer this question. According to this theory, the universe began with an initial spacetime singularity. By "singularity" we mean that the properties of spacetime had become completely indeterminate and causal relationships no longer applied. Hence,

*cf. L. Z. Fang and Y. Q. Chu "From Newton's Laws to Einstein's Relativity", World Scientific & Science Press, 1987.

singularity can tell us nothing as to how the universe ended up with the notion of space, time and vacuum during the inflationary stage.

The origin of singularity itself is even further outside classical cosmology. The founder of the Big Bang, Gamow, quoted the following passage from St. Augustine's "Confessions" to describe his own state of mind when pinned down by the problem of singularity:

> *"Some people talk this and that on what came before God created Heaven and Earth. For such people who dare to pursue such deep propositions, God has prepared Hell".*

This passage was cited by Gamow in an article written in the fifties.

Now, after thirty years, in the early eighties, people can glimpse an outline of the exit from Hell.

The Exit: "Being from Non-Being"

The first move, or the beginning of the universe, has always been regarded as exceeding the bounds of physics, for it seems impossible in principle to determine the first move by physical means.

First, we cannot look for its cause before the first move or the beginning of the universe. For the concept of the first move or beginning of the universe implies that nothing existed before.

Next, we cannot look for the cause of the first move or the beginning of the universe outside the universe. For the concept of universe implies that nothing exists outside.

Thirdly, we cannot go inside the universe either to look for the cause. For the concept of the first move or the beginning of the universe implies that all things in the universe are its results.

Thus, we can go neither before the genesis of the universe, nor after the genesis, nor outside the universe to look for the physical basis of the first move. The conclusion can only be:

Nothing initiated the first move or the beginning of the universe.

Superficially, we have come to a dead end! But the exit is precisely in this dead end. The so-called "nothing" means "nonbeing", hence, the last statement can be written equivalently as:

"Non-being" can initiate the first move or the beginning of the world.

Thus, the solution of the problem of the first move is discovering how the universe can be generated from non-being.

"Being from non-being", how can this be? This proposition seems to be a purely metaphysical one; how is it possible to build a physical theory of

Fig. 11.1. A flag of "NOTHING". According to Taoism, being came from non-being.

"being from non-being"? People have had this misgiving for a long time, and it is a justifiable misgiving.

In 1982, the situation began to change. That year, a symposium on "The Very Early Universe" was held in Cambridge, England, in which, for the first time, the genesis of the universe from nothing was studied as a physical problem. Soviet physicist Linde said a few things at the Symposium which were representative of many a cosmologist's attitude on "being from non-being". He said:

> *"The possibility that the universe was generated from nothing is very interesting and should be further studied. A most perplexing question relating to the singularity is this: what preceded the genesis of the universe? This question appears to be absolutely metaphysical, but our experience with metaphysics tells us that such metaphysical questions are sometimes given answers by physics".*

Linde's comment was not made on the spur of the moment, but had the entire development of cosmology as a background. In particular, several "key answers given by physics" to the question of "being from non-being" had been noted.

One of the keys is knowing what "non-being" is. According to modern cosmology, "non-being" is described as follows:

"What is outside the universe is non-being" or "Nothing can exist outside the universe".

This is a very strong statement. In logic, it is known as a *categorical negative*. Many precedents in physics tell us that the most basic physical laws are often stated in the form of *categorical negatives*. For example the First Law of Thermodynamics states that a perpetual motion machine of the first kind is impossible; the Second Law of Thermodynamics states that a perpetual motion machine of the second kind is impossible; the Principle of Relativity states that it is impossible to measure absolute velocity; the Principle of Equivalence states that it is impossible to distinguish locally inertial mass and gravitational mass. These *categorical negatives* possess a great richness of physical content. From them we can obtain many affirmative, physical conclusions.

The physical theory of "being from non-being" has also opted for this method to deal with the question of cosmic genesis. That is, the framework of this theory is this: starting from the negative statement "nothing exists outside the universe", we shall give affirmative conclusions regarding the first move. Specifically, with "what is outside the universe is non-being" as a premise, we shall attempt to determine the time, space and vacuum of the inflationary period.

Without Time, Without Space

Another key reason why the physics of "being from non-being" has developed lies in our recognition of the notion that time and space themselves have a limited range of application.

Many theories of ontology maintain that time and space are the most indispensible contents or forms of existence, that is, existence and space-time are necessarily connected. However, modern physics tells us that there is not enough basis for this "indispensability". On the contrary, relativity and quantum theory have strongly hinted that, under certain conditions, the concepts themselves of time and space will lose all meaning.

According to relativity, an observer in relative motion will experience different time values. The most vivid illustration is the story of the twins: two identical observers who started together, ended together, but underwent different motions in between would have measured different times for their journeys. Therefore, in order to accurately measure time, the orbit of the measurer's motion must be precisely known. But, on the other hand, quantum theory states that it is impossible to know precisely the orbit of any object in space, otherwise it will contradict the Uncertainty Principle. Thus, in principle, the possibility of a precise time measurement is excluded. The limit on precision is

$$t_P \sim (hG/c^5)^{\frac{1}{2}} \sim 10^{-44}\,\mathrm{s}\ .$$

Here h is Planck's constant. The limit 't_p' states that it is impossible for us to manufacture, by any means whatsoever, a "clock" that can measure time shorter than t_p.

Measurement in space is closely connected with measurement in time. Likewise, we can never find a "ruler" that can measure lengths shorter than l_p, defined by

$$l_p \sim (hG/c^3)^{\frac{1}{2}} \sim 10^{-33}\,\mathrm{cm}\ .$$

The constants t_p and l_p are known as the *Planck time* and the *Planck length*, respectively.

A quantity that cannot be measured in principle is physically meaningless. Hence, it is only in the range above the Planck time and the Planck length that we can use the concepts of time and space. In the range below t_p or l_p, the concepts of time and space are no longer valid. It is a world without time, without space.

A thing that has a limit probably also has an origin. Time and space are concepts with limits, so they may also have an origin. Time should originate in a timeless state and space should originate in a spaceless state.

Without time, concepts such as past, present, or future would be meaningless, for these are all concepts that are attached to time. Einstein seemed to have realized this point earlier on; although he never mentioned it in his scientific papers, his memorial article at the death of his good friend Besso contained the following passage:

> *"Now, although he has left this odd world a little before I, this is nothing, for we who believe in physics all know that the distinction between past, present and future is only an obstinately held illusion".*

If there is no distinction between the past, the present and the future, then we cannot define "first" or "beginning", since "first" and "beginning" are all concepts that are attached to time. We therefore see, once we determine how time originated in the timeless, that this "origin" will be the physical basis for the "first move" or the "beginning of the universe". Thus, we have also avoided using singularity as the starting point of the universe.

A Self-Contained Framework

The possibility that time has an origin has given us a deeper understanding of "being from non-being". This proposition is often understood as a past "non-being" producing the present "being". Such an understanding is adequate for local systems. For example, the past had no solar system and the present now has generated a solar system; the past had no mankind and the present has generated mankind. However, this understanding is incorrect when applied to the entire universe. For the implication of "past non-being" is not equal to "not anything"; the word "past" itself implies time already exists. The "being from non-being" theory of cosmic genesis does not say that there was no universe in the past, and that now there is a created universe — the error here is that it tacitly assumes that time is something that can exist outside the universe.

If we accept that the "past non-being produced the present being", then we shall not be able to understand why Laotzu simultaneously maintains that "Tao generates One, One generates Two, Two generates Three, Three generates all things", for then, "Tao generates One" and "Non-being generates being" would seem quite incompatible. On the other hand, if we use the following framework to understand "being from non-being", then it will be in complete harmony with "Tao generates One". This framework is:

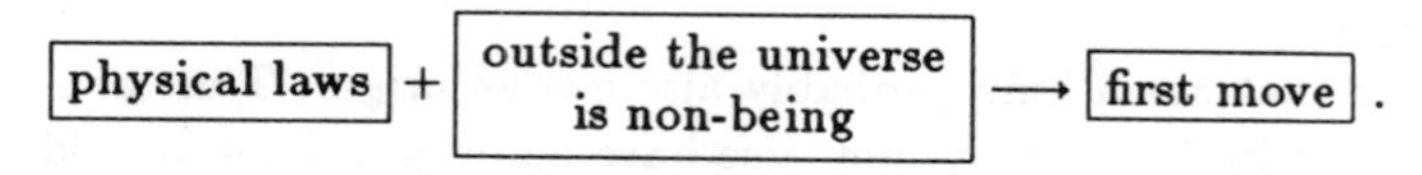

The logic of this framework is this: "Why is the matter of the universe in such a state of motion? Because all this matter is not outside the universe". It is precisely the non-being outside the universe that determines the being in the universe. This is the essence of "being from non-being".

For comparison, the following is the framework of usual physical problems:

physical laws + initial conditions ⟶ consequences .

This framework embodies causality. The initial conditions express the cause, which determines the subsequent events or the effect.

The most salient difference between the two frameworks is this: the "being from non-being" framework involves no other conditions. From physical laws alone, the original state of the universe is uniquely determined, whereas the cause-and-effect framework requires, besides physical laws, the knowledge of initial conditions. Hence the "being from non-being" framework is a self-sufficient framework, while the cause-and-effect framework is not.

The Wave Function of the Universe

Hawking and others of the University of Cambridge, England, were the first to build a logically complete theory of "first move" according to the self-sufficient model.

The first key point in Hawking's theory is the development of a "timeless" physical theory. It states that a timeless space-time is fundamental and that all fundamental physical laws should be written in this "space-time". The characteristic feature of this basic "spacetime" is the impossibility of prescribing a time coordinate; it is essentially Euclidean. Only after applying an analytical continuation of the theoretical results in this Euclidean space can a time coordinate be defined, and comparisons made with observations involving time. That is, time is not the most basic physical quantity; rather, it is only an observational idea. Einstein realized that:

> *Space and time are modes of our thinking and not conditions of our life.*

An important result of this theory is that, for some physical solutions, we cannot obtain time by analytical continuation. This means that, under such conditions, description in terms of the time idea is impossible. The concept of time does not hold. The passage from the essentially non-temporal situation to the situation where time can be used quantitatively prescribes the origin of time, that is, how time originated in a non-temporal situation, or how time itself had its "being from non-being".

The second key point of Hawking's theory is the writing down of the wave function of the universe according to the condition, "outside the universe is non-being". According to quantum theory, all the properties of a system are described by its wave function. The properties of the entire universe should be described by the wave function of the universe. Since time is not a basic quantity, there can be no "non-stationary" wave functions, only "stationary" wave functions in a timeless physical theory. For example, the base state wave function is the most important stationary wave function. Thus, the key question is to find the base state wave function of the universe.

Generally speaking, when calculating the base state wave function of a system, we need to know the system's environment and history, that is, the relevant boundary conditions. For a timeless theory, there is, of course, no question of "history". But boundary conditions are still required. To find the wave function of the universe, we must know its boundary conditions. What are the universe's boundary conditions? According to the self-sufficient assertion, "outside the universe is non-being", we know that the universe's boundary conditions are:

The universe has no boundary.

The wave function determined by this condition is the wave function of the universe.

Once the wave function of the universe has been determined, we can study the various origins. The first origin is the origin of time. On analytically continuing the wave function of the universe into the time representation, we shall find that the concept of time can only be used over a certain range. The point where applicability of time meets inapplicability of time is the origin of cosmic time. Thus, the physical state of the universe at this point is just a "first move". This is a self-contained theory of the "first move".

This self-contained, theoretical system of Hawking is inspiring. But using it to solve the wave function of the universe is certainly no easy matter. At present, quantitative calculations have been made only for several highly simplified cases, and none of these can be regarded as a solution for the real universe, all being toy models. Nevertheless, these toys are valuable in that they show that quantitative calculation of the "first move" is at least, in principle, possible.

The "Hardest" Thing

It is obvious that physics is not content with toy models, but instead wants a theory describing the real universe. To be precise the theory should not only be aesthetic, but also include results of observable tests.

How does one test the theory of the "first move"?

As mentioned in previous chapters, cosmology is like archaeology. The latter tries to deduce the development of human society from pre-historical remains. In cosmology, our aim is to deduce the evolution of the universe from cosmic remains.

Ideally, the available remains should be in their original or as close to their original condition as possible. Things, which have been disturbed, no longer contain historical clues. This is why archaeologists are always interested in ancient tombs. From sacrificial objects one can effectively determine the social situation at the time when the tomb was constructed.

Similarly, cosmologists are interested in digging cosmological "tombs" from which they will be able to look back into the early universe. Up to now, the cosmological "tombs" studied include: objects with large redshifts, microwave background radiation, chemical elements, and matter-antimatter. All those are remnants of various eras. Furthermore, the things that are more difficult to change are the remains from an earlier era. For instance, elements are more difficult to change than galaxies, so elements are remnants of an era earlier than galaxies. Particles and antiparticles change even slower than elements, so particles are probably remnants of an era even earlier than chemical elements. Thus the remains from the creation of the universe should be the hardest things, or, in other words, the thing least conducive to change.

The thing which is most difficult to change is one that is very familiar to us — it is space-time. In Newton's mechanics space and time are absolute and unchangeable. Einstein's general relativity forsakes the fixity of space-time and shows that space-time is curved due to the existence of matter. However, as with most problems in physics, space-time is considered a fixed platform and the effects of general relativity can be neglected. This is the same as assuming that space-time is the hardest of all physical entities (Fig. 11.2).

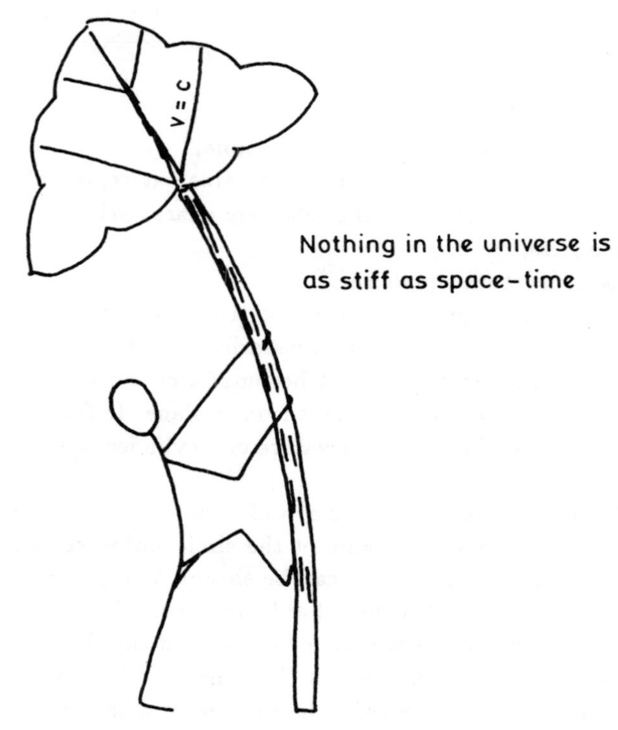

V = C
Nothing in the universe is
as stiff as space-time

Fig. 11.2. Space-time is the hardest thing among all physical entities.

Therefore, the features of space-time themselves must have been formed in an era earlier than all other remains. Namely, space-time should directly be determined by the "first move". So one experiment testing the theory of the "first move" is to search whether the theory can consist with the observable feature of space-time — the topology of cosmic space-time. All above-mentioned statements are summarized in Table 11.1.

Table 11.1. Cosmic remains of various eras.

Remains	Age of the universe
Topology of spacetime	10^{-44} seconds
Asymmetry of particle & antiparticle	10^{-36} seconds
Abundances of elements	3 minutes
Microwave background radiation	10^5 years
Objects with large redshift	$\sim 10^{10}$ years ago

Topology of Space-Time

Topology is a global property of space-time.

Let us consider a piece of paper on a table. It represents a part of an infinite plane. Such a plane has the following characteristics:

1. boundaryless,
2. flatness (curvature equals zero).

However, the contrary proposition is not true, namely, a geometry that is flat and has no boundaries is not necessarily an infinite plane. If we glue opposite sides of a paper together, it becomes a cylinder (see Fig. 11.3). At first glance, a cylinder is quite different from a plane. In fact, it can be shown that a cylinder is also flat or the curvature of a cylinder is equal to zero at all points.

The difference between a cylinder and a plane is not local, but global. Let us draw a circle on a plane and let the circle shrink continuously on the plane. It is obvious that the circle can be shrunk to a point. Such geometry is called *simply connected.* On the other hand, for a cylinder, there are circles (see Fig. 11.3) which would not shrink to a point if the shrinking is continual and the curve is to be maintained on the cylinder. Such geometry is called *multiply connected.* In other words, the difference in the connection between the plane and the cylinder is due to a difference in topology.

The topology of cosmic 4-dimensional space-time is of concern to cosmology.

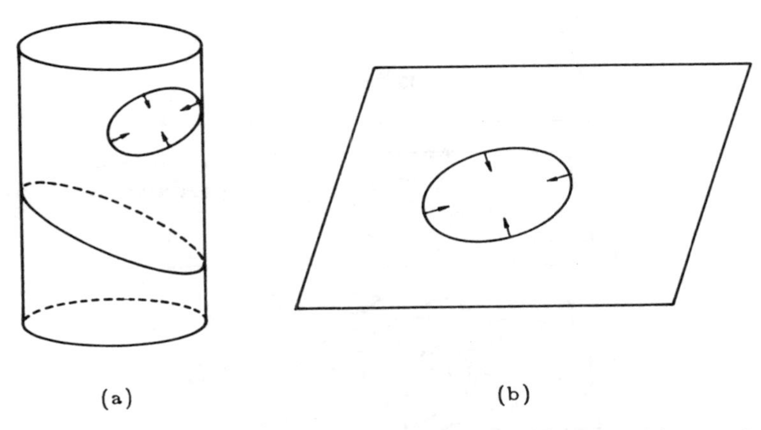

Fig. 11.3. (a) A cylinder or radius R appears to be a curved surface. (b) It can be shown that the cylinder is as flat as a plane. However, a cylinder is not the same as a plane in all aspects. For example, among all the geodesics through any point on a cylinder, one will be closed. This does not happen on a plane.

A Multiply Connected Universe?

In the ordinary scales space and time are like a big plane; they are simply connected. However, on a small scale, such as Planck length and Planck time, space-time is probably multiply connected. (Fig. 11.4).

We have already mentioned that no space and no time can be used in a scale smaller than the Planck scale. In this case "present", "past", and "future" will lose meaning, "up-down", "front-behind", "left-right" will also lose meaning. It seems very hard to imagine a world without the notions of up-down and before-after. In fact, it is not too difficult to understand. For instance, let us consider two runners A and B on a circular runway. If you do not have all the information from their starting point to the present you would not, in general, be able to tell who is leading the race, because the track is circular. Seen from one half of the course A is in the lead, while from the other half B seems to be in the lead. This means that the concept of front-behind is invalid. If space-time, like a racetrack, is circular, the order given by space-time will then also be invalid.

In standard problems in physics, the scales are very different from Planck's scale and the concept of the order given by Planck's scale and space-time can be used. However, during the creation of the universe, or the Planck era, the problems were different. During that time the scale of the universe as a whole was as small as Planck's scale. It can be seen that the universe as a whole is microcosmic. The evolution of the topology of cosmic space-time became the

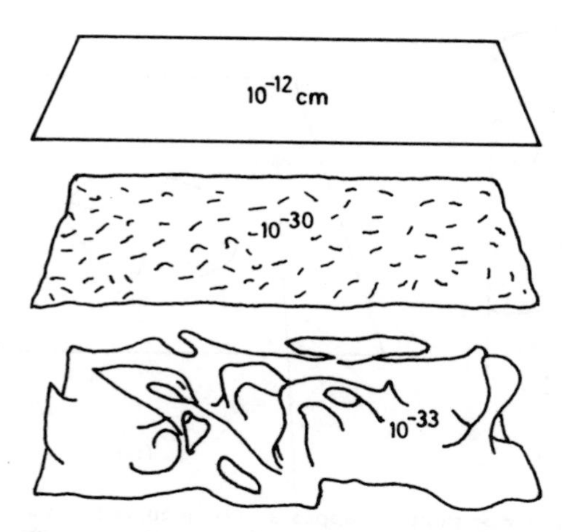

Fig. 11.4. (a) When the ocean is seen by an observer flying 5 km above it, its surface appears flat. This is similar to space-time appearing flat at scales larger than 10^{-12} cm. (b) When the observer descends to 100 m, he sees waves. Similarily, when we approach the scale of 10^{-30} cm, space-time is also waved. (c) Finally, when the observer drops into and floats on the ocean with a lifejacket, he sees the waves breaking into foam which corresponds to the foam-like structures in a fluctuating space-time at Planck scales.

main physical process. Namely, the "first move" is first to "move" the cosmic space-time, to determine the topology of the cosmic space-time.

After the expansion of the universe, the scale of cosmic energy decreased, the cosmic temperature also decreased, and space-time became harder and harder. Since then, no physical processes were able to disturb the space-time topology, so the topology become unchangeable and remains so until the present. Because the space-time topology in the Planck era is probably multiply connected, today's cosmic space-time topology will probably also be multiply connected.

This conjecture seems to be contrary to ordinary experiences. Multiple connectivity implies the invalidation of "up-down" and "before-after", while ordinary experience tells us that such concepts are quite useful. This contradiction can be explained easily. Let us once again consider the race between A and B. In a dash or a short race, it is not difficult to define the order of A and B. Because a dash uses only part of a circular racetrack, no confusion related to the topology of the racetrack can occur. Ordinary life related only to scales much smaller than that of the universe is like a dash, so the multiple connections of the universe as a whole would not lead to the confusion of "before-after" in the usual experience.

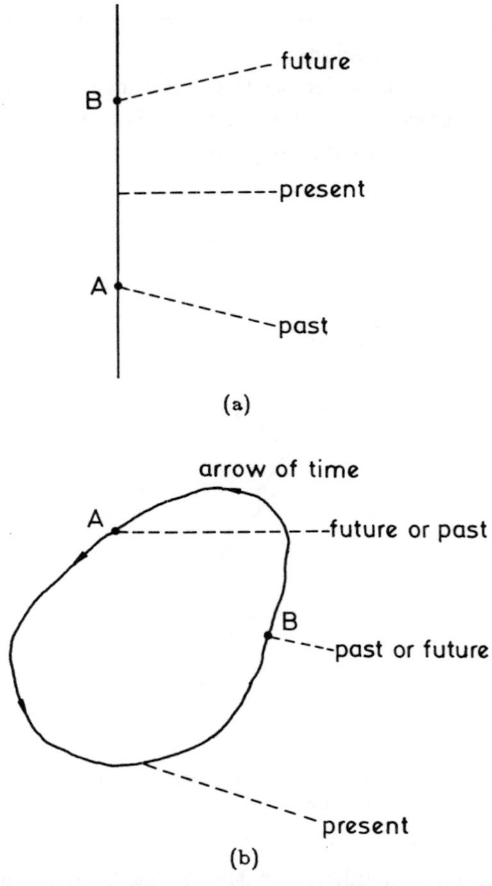

Fig. 11.5. (a) It is possible to define the order of A and B by using the concepts of past, present and future, if A and B are on a line. (b) It is impossible to do that on a circle.

If we want to show the multiple connectivity of the universe, we should search for cosmic "long distance runners", i.e. large scale structures. Indeed, several observed results seem to show that the topology of the universe may be multiply connected.

The first possible evidence for the multiple connectivity of the cosmological space comes from the distribution of a quasar's number with respect to its redshift. Quasars possess larger redshifts, mainly in the range of 0.1 to 3. In this range the distribution of a quasar's number is not uniform. For some redshift values a quasar's number is greater, for others the number is less.

Since the 1970's the distribution of a quasar's number with respect to its redshifts was found to be periodic.

The redshifts of quasars denote their distance from us. Therefore, the periodicity in the quasar's distribution may show that the distribution of a quasar's number against distance is periodic, as shown in Fig. 11.6.

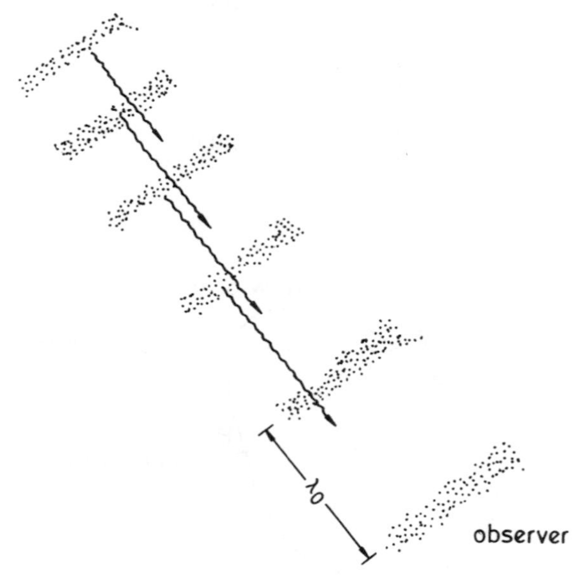

Fig. 11.6. The periodicity in the redshift distribution of quasars may be due to the periodicity in a quasar's spatial distribution.

Why is there such periodicity in the quasar's distribution? A possible explanation of this periodicity is the ***multiply connected universe***. A multiply-connected space has closed "racetracks" on large scales. Long distance runners show periodic motion due to the circular racetrack. In a similar way we will find the periodic distribution of objects.

For instance, a two-dimensional torus is obtained from a flat simply-connected plane by identifying point (x, y) with points $(x + ma, y + nb)$, where a, b are constants and m, n are all integers (Fig. 11.7). An observer in such a torus will find that the observed picture is the same as a plane, but the distribution of matter is periodic with "wavelength" a in the x direction and b in the y direction, and the picture looks like a plane lattice (see Fig. 11.7).

The second possible evidence for multiple connectivity comes from the association between quasars and galaxies. Galaxies are objects with smaller

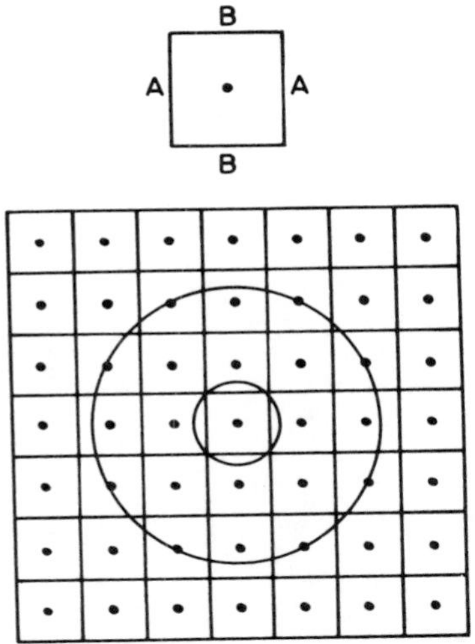

Fig. 11.7. (a) A two-dimensional torus can be constructed by identifying the point (x, y) with the points $(x + ma, y + nb)$, where a, b are constants and m, n are integers. (b) An observer on such a torus will find that the distribution of matter is periodic, just like a plane lattice with "wavelength" a in the x-direction and b in the y-direction.

redshifts; they are nearby. Physical features of galaxies are also quite different from quasars. Therefore, the distributions of galaxies and quasars should be independent of each other. However, statistical analyses showed a positive correlation between the distributions of galaxies and quasars. It is very difficult to explain why objects with such different redshifts are correlated in position. A possible solution lies in the assumption of multiple connectivity. Galaxies and quasars may just be manifestations of the same spatial object at different stages of its evolution, both visible simultaneously because of the multiple connectivity of space.

All evidence mentioned above is only tentative. The significance of this research lies mainly in methodology. It shows that the creation of the universe or "the first move" can also be studied as follows:

1. In *observational cosmology*, to determine the space-time topology of the universe as a whole by means of analysing the large scale structure of the distribution of objects, such as quasars and galaxies.

2. In *theoretical cosmology*, to construct a model of the creation of the universe which can explain why the space-time topology is what it is.

Chapter Twelve

THE ANTHROPIC PRINCIPLE AND PHYSICAL CONSTANTS

Loopholes in the Cosmological Framework

This is the last chapter and we are now in a position to take a panoramic view of the development of cosmology.

Figure 12.1 summarizes the entire field of cosmology. We may even describe it as a framework of the epistemology of the whole of physics. Each block in the figure represents a question. Arrows point to the various questions among logical relations. For example,

Question: Why should the future be like this and not like that?
Answer: Because the physical laws are like this and the present is like this and not like that.
Q: Why was the past that and not something else?
A: Because the physical laws are this and the first move was this and not that.
Q: Why was the first move this and not that?
A: Because outside the universe is non-being and the physical laws are this.
Q: Why is "outside the universe" non-being?
A: This is a self-contained assertion. It is its own basis.

We can see that, as far as this framework is concerned, only self-contained assertions can serve as terminal points, otherwise, we can always enlarge the framework by asking another "why".

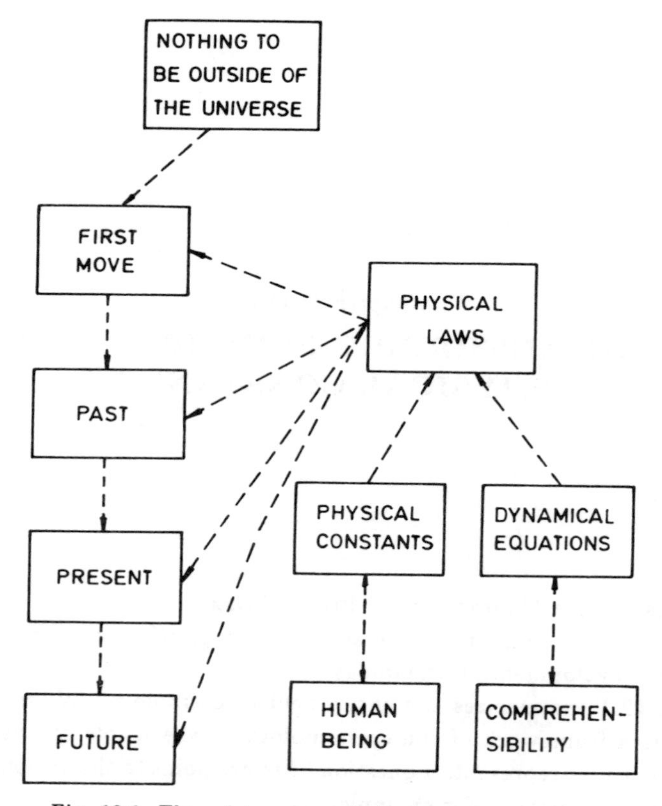

Fig. 12.1. The epistemology of physics and cosmology.

The other terminal point in the diagram is physical laws. Obviously, physical laws are not self-contained. They cannot serve as their own basis. Therefore, we can ask, why are physical laws this and not that?

Physical laws have two kinds of content, one is the form of the dynamical equation, the others are the values of the constants. Why are dynamical equations this and not that? Why do physical constants have these values and not some other ones?

If we judge the framework shown in Fig. 12.1 using the criterion of completeness, then these two questions are its gaps.

Obviously, a thorough solution of these two problems cannot be accomplished by finding the law of these laws, because for any law of laws, we can still ask why *this* and *not that*. That is, the law is still not self-contained.

The Anthropic Principle

One way of filling the gaps in Fig. 12.1 is to adopt the Anthropic Principle.

The basic idea of the Anthropic Principle is this: the reason why the universe is so is because, if it is not so, then there will be no mankind, and nobody to study the universe. In brief, things present are so because man exists at present.

This logic is very different from the familiar mechanistic view. One of the basic tenets of the mechanistic view is that the existence of the universe and the existence of human beings in the universe are unrelated.

An example will clarify. If we ask, why is the universe so big, the answers based on *the mechanistic view* will go as follows:

The universe began with the big bang;

The size of the universe can be indicated by multiplying the age by the speed of light;

The age of the universe is about 10^{10} years;

Hence the size of the universe is at least 10^{10} light years.

The answers based on *the Anthropic Principle* will take the following form:

It is man who asks this question;

Man is a living being;

Life requires carbon;

Carbon is generated in the stars;

Supernova explosions scatter carbon through space;

It is only after nuclear burning that stars can undergo supernova explosions;

Nuclear burning takes more than 10^{10} years;

Hence the size of the universe is at least 10^{10} light years.

Obviously, wherever we can use the mechanistic view to explain the issues, we do not need nor wish to re-explain them by using the Anthropic Principle. For the Anthropic Principle depends on the knowledge of man and our knowledge of man is rudimentary. Also, with the Anthropic Principle, it is difficult to make any verifiable predictions. On the other hand, the whole of physics after Galileo and Newton can be said to be based on the mechanistic view, and every achievement in physics is a demonstration of the effectiveness of the mechanistic view.

Nonetheless, regarding these two questions, the mechanistic view is completely useless: why do the basic physical constants have these values? Why do the basic dynamical equations take these forms? The Anthropic Principle may provide some explanations for these two questions, although they are very general — that is why we used double-headed arrows in the diagram (Fig. 12.1).

We now discuss separately how the Anthropic Principle is used to explain the physical constants and the dynamical laws.

Basic Physical Constants

There are only few basic physical constants. These are the more familiar ones:

$$
\begin{aligned}
c &= 2.997\,924\,58 \times 10^{10} \mathrm{cm\,s^{-1}} \\
h &= 1.054\,589 \times 10^{-27} \mathrm{g\,cm^2\,s^{-1}} \\
G &= 6.673 \times 10^{-8} \mathrm{cm^3\,g^{-1}\,s^{-2}} \\
e &= 4.803\,24 \times 10^{-10} \mathrm{e.s.u.} \\
m_e &= 9.109\,53 \times 10^{-28} \mathrm{g} \\
m_p &= 1.672\,61 \times 10^{-24} \mathrm{g}
\end{aligned}
$$

To be included among the basic physical constants are the interaction constants of weak and strong interactions.

The scales of all things in the universe can be formed by combining these basic physical constants. The frequently used ones are,

$$
\begin{aligned}
\text{size of atoms } a_0 &= h^2/m_e e^2 & &\sim 10^{-8}\,\mathrm{cm} \\
\text{atomic density } p_0 &= m_p/a_0^3 & &\sim 1\,\mathrm{g\,cm^{-3}} \\
\text{size of nuclei } r_p &= h/m_p c & &\sim 10^{-13}\,\mathrm{cm} \\
\text{nuclear density } p_N &= m_p/r_p^3 & &\sim 10^{15}\,\mathrm{g\,cm^{-3}} \\
\text{ionization energy } E_0 &= e^4 m_e/h^2 & &\sim 10\,\mathrm{eV} \\
\text{energy of molecular interaction } E_m &= E_0(m_e/m_p) & &\sim 10^{-2}\,\mathrm{eV}
\end{aligned}
$$

We now proceed to analyse what requirements we must place on these constants in order that man can exist.

Necessary Conditions for the Existence of Homo Sapiens (I)

As a first step, let us discuss some basic properties of the Earth, since the Earth is the foremost condition for our existence.

All matter on Earth is in a state of equilibrium between electrostatic attraction and degenerate electron pressure. If the number density of electrons in this state is n, the distance between electrons is about

$$\Delta x \sim n^{-\frac{1}{3}} . \tag{12.1}$$

Hence the electrostatic energy between the particles is

$$e^2/\Delta x \sim e^2 n^{\frac{1}{3}} . \tag{12.2}$$

On the other hand, according to the Uncertainty Principle, the momentum of the electron is about $\Delta p \sim h/\Delta x$; hence the electron degeneracy energy is about

$$\begin{aligned}(1/2m_e)(\Delta p)^2 &\sim (h^2/2m_e)(1/\Delta x)^2 \\ &\sim (h^2/2m_e)n^{\frac{2}{3}} .\end{aligned} \tag{12.3}$$

Thus, for equilibrium between electrostatic energy and electron degeneracy energy, we have

$$e^2 n^{\frac{1}{3}} \sim (h^2/2m_e)n^{\frac{2}{3}} , \tag{12.4}$$

or

$$n \sim e^6 m_e^3/h^6 .$$

The density of matter on Earth so found is on the order of

$$p \sim nm_p \sim e^6 m_e^3 m_p/h^6 \sim 1 \text{ g cm}^{-3} . \tag{12.5}$$

This result agrees well with experimental values.

From (12.5) we can also find the relation between the Earth's radius R and its mass $M_\oplus$:

$$R \sim (M_\oplus/p)^{\frac{1}{3}} \sim (M_\oplus/m_p)^{\frac{1}{3}}(h^2/m_e e^2) . \tag{12.6}$$

Using this equation, we can now calculate what the Earth's mass should be in order that humans may exist.

Necessary Conditions for the Existence of Homo Sapiens (II)

The human being is kept alive by constant metabolism. Biochemical reactions are constantly taking place in the human body. Such reactions are carried out effectively only under a certain temperature. Too high a temperature will degrade the enzymes and shorten human lives; too low a temperature will halt the reactions and life will be extinguished. The suitable temperature should be one where the heat energy is close to the energy of molecular interaction, that is, the Earth's environmental temperature T should be

$$T \sim E_m/k , \tag{12.7}$$

or

$$T \sim e^4 m_e^2 / kh^2 m_p \sim 100°\text{K} , \quad (12.8)$$

which is indeed the temperature on Earth.

Then, there are three states of matter in the human body, the solid and liquid states in the body, and the human must also breathe gas. Therefore, the Earth must have an atmosphere at a temperature of about 100°K. This requirement places a constraint on the Earth's mass; it cannot be too small. Bodies with small mass have small velocities of escape, and gas would not be retained and would escape into space. This is the reason why asteroids and the Moon do not have an atmosphere. Due to this constraint, the Earth's mass must be large enough to satisfy the relation

$$GM_{\oplus} m_p / R > KT . \quad (12.9)$$

Using (12.6) and (12.8), this can be rewritten as

$$M_{\oplus} > e^3 m_e^{\frac{3}{2}} / G^{\frac{3}{2}} m_p^{\frac{7}{2}} \sim 4 \times 10^{25}\ g . \quad (12.10)$$

This is one condition the Earth's mass must satisfy for humans to exist on Earth.

Fig. 12.2. If the Earth's mass were smaller, then molecules in the air would escape from the Earth and no atmosphere would be maintained on the surface of the Earth.

Necessary Conditions for the Existence of Homo Sapiens (III)

The living human must move and when moving, the human body is constantly subject to forces. Hence, in order to maintain the human body as an integral whole, it must have enough rigidity to withstand impacts, otherwise the body may be broken.

Fig. 12.3. When discussing the necessary condition of the existence of man, we simplify a man to a sphere with radius $H/2$.

Let us simplify the human body into a sphere of diameter H. If the number of atoms in unit volume is n, then approximately $n^{\frac{2}{3}}$ pairs of atoms will interact from two sides of a unit surface area. Each pair has an interaction energy of about E_0.

When the body is under impact, the area subject to the effect is approximately

$$S \sim H^2 .$$

Hence the energy that keeps the body from breaking is approximately

$$E \sim H^2 n^{\frac{2}{3}} E_0 .$$

Usually when a person walks or runs, the scale of energy involved is about one twentieth of the potential energy of the gravity of the body, that is

$$\Delta E \sim mgH/20 , \tag{12.11}$$

where m is the mass of the human body, g is the gravity,

$$m \sim H^3 \rho , \tag{12.12}$$

$$g \sim GM_{\oplus}/R^2 , \tag{12.13}$$

and ρ the body's density.

In order for the body not to break too easily, we require

$$\Delta E < E , \tag{12.14}$$

or

$$\frac{M_\oplus}{R^2} < \frac{20n^{\frac{2}{3}}}{\rho H^2}\frac{E_0}{G} \, . \tag{12.15}$$

Using (12.6) and (12.8), this becomes

$$M_\oplus < \frac{20^3 n^2}{\rho^2 H^6} \cdot \frac{h^6}{G^3 m_e^3 m_p^2} \, . \tag{12.16}$$

The matter of the human body has a density of $\rho \sim 1\,\mathrm{g\,cm^{-3}}$. The number density of atoms is $n \sim \rho/m_p \sim 1/m_p$. The height of the person can be taken as $h \sim 100$ cm. With these values, the last formula gives

$$M_\oplus < 10^{28}\,\mathrm{g} \, . \tag{12.17}$$

This is another constraint on the mass of the Earth in order that human beings may exist on it.

Fig. 12.4. From the conditions (12.11) to (12.15) we can explain why a more massive animal has greater difficulties with locomotion: a dinosaur may only be a reptile; man can walk and run; fleas always jump.

Anthropicism of Physical Constants

We combine the two constraints (12.10) and (12.17) on the Earth's mass:

$$10^{25}\,\mathrm{g} < M_\oplus < 10^{28}\,\mathrm{g} \, . \tag{12.18}$$

The actual mass of the Earth indeed falls within this range, in fact,

$$M_\oplus = 6 \times 10^{27}\,\mathrm{g} \, . \tag{12.19}$$

Let us take another look at the constraints (12.10) and (12.17). Note that they are determined entirely by physical constants. No mass $M_\oplus$ can satisfy the two constraints simultaneously. Therefore man can only exist in a universe with some specified constant values, and the possibility of man's existence cannot be separated from a certain fine-tuning of physical constants. This is the anthropic interpretation of physical constants.

We now proceed to present more evidence for this interpretation.

First, the gravitation constant G.

The ambient temperature on Earth depends on the Sun for its maintenance, and the Sun is at present located in the stable, main-sequence stage (see Chapter 3). It has taken more than one billion years to go from the most primitive life form to humans. Hence, for mankind to evolve, we require the Sun to be well settled in the main sequence stage for several billion years or more. This requirement makes us realize that G must not be too large, nor too small. It can only take the observed value. If G is too large, then the Sun would have soon evolved into the high temperature stage and would not have been able to remain in the main sequence for some billions of years. If G is too small, then it would be impossible for the Sun to reach the main sequence stage and the Sun would still be lingering about in the low temperature stage. Thus only when G takes the present, known value will the solar system be maintained long enough in a condition favorable to the ascent of mankind.

Second, let us discuss the weak interaction constant g_W.

The theory of element synthesis was presented in detail in Chapter 8. The process of element synthesis depends heavily on g_W. If g_W is too large, then neutron decay would be faster, and neutrons would have disappeared through decay at the start of nucleosynthesis. Such a universe only has protons and no neutrons. No heavy elements can form, nor can mankind. On the other hand, if g_W is too small, then the following process would be too sluggish

$$p + \bar{\nu}_e \rightleftharpoons n + e^+ \ .$$

The number ratio between neutrons and protons would have "frozen-out" at a very early cosmic time, when the temperature T was very high: $kT >> (m_n - m_p)c^2$. Since:

$$n_n/n_p \sim \exp[(m_n - m_p)c^2/kT] \ ,$$

we have $n_n \sim n_p$. In such a universe, all neutrons and protons would combine into helium nuclei, and no protons would be left over. The result would be a hydrogen-less universe, once again impossible for the evolution of the human. Therefore, in order to provide various chemical elements that the human needs, g_W must have taken roughly its present value.

Third, let us talk about the parameters of the Grand Unified Theory.

The Grand Unified Theory predicts that protons are unstable, and that they can spontaneously decay into electrons and π-mesons. The proton lifetime is related to the parameters of the theory. Let the average mass of a person be 50 kilograms, then the total number of nuclei in the body is

$$N = 50 \times 10^3/1.7 \times 10^{-24} \sim 3 \times 10^{28} .$$

The corresponding number of protons is about

$$N_p = N/2 .$$

The decay of protons inside the human body means that the body is subject to radiation. The largest dose of irradiation a person can tolerate is one microcurie, that is, not more than ten thousand proton decay events every second. From this we derive a lower limit for the lifetime of the proton:

$$\tau_p > 10^{28}/10^4 \sim 10^{24}\text{s} \sim 10^{17}\text{yr} .$$

The Grand Unified Theory predicts that $\tau_p > 10^{31}$ yr. This value comfortably satisfies the above requirement.

Physical Constants are Constants

For the anthropic interpretation of physical constants above, a number of simplified calculations were made. All the calculations contain a tacit assumption, namely, that all these physical constants are in fact constant. Without this assumption, many of the conclusions would be doubtful.

It is true that at least in experimental physics, the various constants used above can be regarded as invariable constants. But when discussing cosmology, we often deal with billions or even tens of billions of years. Over such vast periods, can these constants still be constant? This question must be answered. At least, we would like to be able to assert that, over the twenty billion years of the universe, physical constants have not changed appreciably, so as to guarantee the accuracy of anthropic calculations.

Therefore, we need to find the upper limit of the rate of variation of various physical constants. We shall discuss these one by one.

If the gravitational constant G varies with time, then the motion of the planets will be affected. When G increases, gravitation becomes stronger. The planets will be closer to the Sun and their periods of revolution will be shorter. If, on the contrary, G decreases, then the planetary periods will be longer. Variations in the periods can be measured by the atomic clock. For Mercury and Venus, the observed result shows that the ratio between the yearly change ΔG and G is bounded above by

$$\Delta G/G < 10^{-11}/\text{years} .$$

That is, throughout the entire course of the evolution of the universe, the gravitational constant is indeed a constant.

To assess the rate of change in the fine structure constant $\alpha = e^2/hc$, we can analyze the spectra of distant bodies. The spectral lines of atoms or ions are closely related to the fine structure constant α, and if we compare the spectra of certain atoms or ions in distant bodies with the spectra of similar compositions on Earth, we shall be able to measure changes in α. The oxygen ion spectrum of a radio galaxy with a redshift $z = 0.2$ and the silicon spectrum of a quasar at $z = 1.95$ have been analyzed, and they gave exactly the same value of α as on Earth.

If the electron charge e varies with time, then this will affect the electromagnetic interaction, as well as the atomic structure and the properties of the atomic nuclei. For example, an isotope of Rhenium, ^{187}Re, is radioactive. It changes by β-decay into ^{187}Os (Osmium-187), with a half-life of about 40 billion years. If e varies, the decay rate of ^{187}Os itself will change. If the yearly rate of decrease of e is one part in 2 billion years (5×10^{-11} yr^{-1}), then the rate of decay of Osmium at the early stage would be so fast that no Osmium would have remained. But there is still much Osmium in the Earth's crust, hence the early decay rate could not have been too fast. This gives an upper limit for the yearly rate of change of e:

$$\Delta e/e < 10^{-13}\ \text{yr}^{-1} .$$

If the electron mass m_e can change, then this will also make ^{187}Os decay faster. Hence the existence of Osmium also gives an upper limit to the yearly rate of change of m_e:

$$\Delta m_e/m_e < 10^{-13}\ \text{yr}^{-1} .$$

Anthropicism of Physical Laws

Lastly, let us discuss how to use the Anthropic Principle to explain why physical laws have this and not that form.

In the anthropic view, physical laws are gradually built up in the course of man's study of cosmic phenomena. Therefore, physical laws must possess the property of being comprehensible to man. Laws that man can understand should conform to logic, for man's understanding proceeds only in logical forms. The requirement "to conform to logic" strongly restricts the form of laws, in other words, not many forms of laws satisfy the requirement "to conform to logic". Einstein said:

> *"What I am really concerned with is, whether or not God can build up this world with different forms, that is to say, whether the requirement of logical simplicity has left any freedom at all".*

Thus, basic laws that satisfy the requirement "comprehensible to man" may leave no room for free choice.

Fig. 12.5. Why is the universe comprehensible?

In short, anthropicism relating to physical laws can be summarized by the following formulae:

Why are the physical laws like this? Because only laws of this form are comprehensible to man. In greater detail:

Physical laws are comprehensible to man;
Comprehensible laws should conform to logic;
Conformity to logic leaves no room for free choice;
Hence, physical laws are determined by "comprehensibility to man".

Einstein also emphasized comprehensibility in the following form:

"The most incomprehensible thing in the universe is that the universe is comprehensible",

We can now answer as follows. The "comprehensibility of the universe" is precisely due to the existence of the human who has the power of comprehension and who wants to comprehend this question. For, if the universe can produce, through evolution, humans that have the power of comprehension, then the universe must be comprehensible. Otherwise, if the universe were incomprehensible, then we would not even be able to begin to talk about the existence of the human with the power of comprehension, or pose the question on why the universe is comprehensible. The existence of the human being supports the idea that humans can only exist in a comprehensible universe. In brief:

> Only "comprehensible" things are comprehensible. Just like "outside the universe is non-being".

This looks like a tautology. However, for cosmology, both are self-contained assertions of great value.

Physics is Tolerant

Cosmology is going through a fascinating period. A question that has always been considered a topic of metaphysics or theology — the creation of the universe — has now become an area of active research in physics. Traditional physical methods apply to the study of cosmology, namely the mutual corroboration of theory and observation.

More than three hundred years ago, Newton insisted on using natural forces to explain celestial motions, which had previously been regarded as divine. But in the face of "primordial motion", he retreated. More than thirty years ago, Gamow championed the evolutionary viewpoint, using the big bang theory to explain the origin of what had been thought to be the eternal chemical elements. But when it came to the origin of the big bang itself, he faltered, resorting to the following passage from the *Confessions* of St. Augustine to express his sense of defeat:

> *I answer him that he asketh what did God do before He made Heaven and Earth? ... He was preparing Hell for prayers into mysteries.*

Physics is tolerant and has so far been unprepared to cast anything into Hell. What we can expect is that with the development of the physics of the creation of the universe, ancient creation myths, medieval church doctrines, and much still current metaphysical wild theorizing will all become historical exhibits, testifying to an earlier stage of human culture.

INDEX